Geology of the American Southwest

A Journey through Two Billion Years of Plate-Tectonic History

The processes of geology form our landscape over immeasurable expanses of time, and nowhere is this more self-evident than in the spectacular scenery of the Southwest of the United States of America. Two billion years of Earth history are represented in the rocks and landscape of this region, creating natural wonders such as the Grand Canyon, Monument Valley, the Black Canyon of the Gunnison, and Death Valley.

The Southwest stretches from southern Utah and Nevada in the north, to Arizona, New Mexico, southeastern California, and northern Mexico to the south. This region is considered a geologist's 'dream,' since its rocks provide a slice through a huge range of Earth history, and provide examples of many of the geologic processes shaping the Earth. For this reason, the region attracts a large number of undergraduate field classes and amateur geologists.

Geology of the American Southwest provides the first concise and accessible account of the geology of the region. It relates local geological events to global plate tectonics, and documents complex episodes of sedimentation, mountain building, and crustal stretching, describing events leading to features of the modern landscape. Its broad, chronological approach differs from other books about the geology of the region, which either focus on smaller regions or are organized by travel route. This book will prove invaluable to students studying the geology of the Southwest. It will also appeal to anyone interested in geology and landscape, and is a valuable guide for visitors to the National Parks and Monuments of the region.

W. SCOTT BALDRIDGE is a Research Scientist in the Earth and Environmental Sciences Division, Los Alamos National Laboratory, and an Adjunct Professor at the University of New Mexico.

T0276221

Geology of the American Southwest

A Journey through Two Billion Years of Plate-Tectonic History

W. SCOTT BALDRIDGE

CAMBRIDGE
UNIVERSITY PRESS

CAMBRIDGE UNIVERSITY PRESS
Cambridge, New York, Melbourne, Madrid, Cape Town, Singapore,
São Paulo, Delhi, Dubai, Tokyo, Mexico City

Cambridge University Press
The Edinburgh Building, Cambridge CB2 8RU, UK

Published in the United States of America by Cambridge University Press, New York

www.cambridge.org
Information on this title: www.cambridge.org/9780521016667

First published 2004
Fourth printing 2007

A catalogue record for this publication is available from the British Library

Library of Congress Cataloguing in Publication Data

Baldridge, W. Scott, 1945–
    Geology of the Southwest : a journey through two billion years of plate-tectonic
history / W. Scott Baldridge.
        p.   cm.
    Includes bibliographical references and index.
    ISBN 0 521 81639 4 (hardcover) – ISBN 0 521 01666 5 (paperback)
    1. Plate tectonics – Southwest, New.   2. Geology – Southwest, New. I. Title.
QE79.5.B35 2003
557.9–dc21   2003051239

ISBN 978-0-521-81639-7 Hardback
ISBN 978-0-521-01666-7 Paperback

Throughout his thirty-four years of teaching undergraduate students at Hamilton College in Clinton, New York, Donald B. Potter sought to instill in them a love for geology and an appreciation for the importance of science in a liberal arts education. With infectious enthusiasm, he encouraged and guided students to deduce geological relationships from empirical observations, using the field as an outdoor laboratory whenever possible. An impeccable field geologist, researcher, and scholar, Don involved undergraduates in research before it was fashionable to do so. Among the lessons taught by Don were the fundamental importance of critical thinking and writing skills. This book, then, is dedicated to Donald Brandrith Potter, whose commitment to a career of teaching and whose passion for geology helped to educate, motivate, and inspire more than a generation of students.

Contents

* These plates are available in colour for download from www.cambridge.org/9780521016667

Boxes

Preface

The geologic history of the American Southwest is both fascinating and important, not least because it is openly revealed to both the professional Earth scientist and casual observer alike. The exposures in this arid to semi-arid region are generally superb. This book, then, is intended to present a systematic and comprehensive picture of the geology of the Southwest since the formation of its earliest rocks in subduction zones, through the formation and fragmentation of at least two and possibly additional supercontinents. It will supplement other books, including the popular road guides, presenting more detailed pictures of the geology of the region.

But the real importance of geological studies of this region lies in their broad application to other areas of the Earth. Geological paradigms developed in the Southwest have global import. Thus, a secondary purpose is to highlight the numerous concepts that have grown out of study and research in the Southwest and West and that have bearing elsewhere in the world. Since the geographical and topographical surveys of the middle and latter part of the nineteenth century, the geology of the West has strongly influenced the development of the geological sciences. Examples derived from the American Southwest have been used in the professional training of geologists for over a hundred years. In part, this book will illustrate how concepts derived from study of modern rocks feed back into an understanding of rocks from earlier periods of Earth's history, and vice versa. Many important global geological concepts have emerged from the Southwest. Some of the geological paradigms developed in this region are in the process of being applied in other, less well exposed and studied, regions of the world.

Because of the vigorous and ongoing research activity in the Southwest, part of the approach used here is to highlight research issues and unsolved problems, and to point out implications for understanding geological problems worldwide. In this manner, the reader may appreciate the ongoing nature of science, which is, after all, a process of exploration. Thus, this approach emphasizes the dynamic, human intellectual endeavor that *is* science and, implicitly at least, the processes by which understanding is arrived at. Presenting some of the limitations of our understanding can be frustrating, because we are reminded of how much is not yet known. At the same time, it is encouraging to be reminded of how much *has* been learned. And major progress continues to be made, as geologists apply new tools and acquire the new data that allow them to choose between alternative hypotheses. Therefore, we can expect that, of the uncertainties highlighted in this book, non-viable hypotheses will eventually be abandoned and successful ones will become accepted, probably in modified form.

The organization of this book differs from that of many devoted to the geological history of parts of the Southwest, or any region for that matter, in two significant respects. First, books on the geologic history of a region typically devote only a few paragraphs or perhaps a chapter to Precambrian time. In diagrammatic time scales, the Precambrian scale is invariably greatly compressed. Yet the Precambrian includes most of geologic time and, in the Southwest, the most important interval of time. For it is during the Precambrian, or more specifically, the Proterozoic (the latter part of the Precambrian) that initial formation of continental crust in the Southwest occurred. Everything else — most of what we see and live on in the Southwest — is built upon this fundament. In the present book, three chapters are invested on the Precambrian. Second, chapters are organized around major geological events that affected the Southwest, rather than by the conventional time-stratigraphic scale, which, when developed in the nineteenth century, reflected events and 'packages' of rock units mainly in Europe. Each chapter of the book begins with a *linear* time scale, with key events referenced.

To minimize interruptions to the text, boxes are used for selected definitions and for detailed discussions of specific features. Some of these are, in a sense, 'mini' papers. In a few instances, figure captions may stand separate from, and augment, discussions in the text. Thus, they constitute 'micro' papers.

This book assumes a familiarity with the terminology and concepts presented in a basic course in physical geology and therefore is directed toward upper level undergraduates and graduate students. Selected terms and concepts that are typically not explained in introductory textbooks on physical geology are discussed in boxes and in the Glossary. It is hoped that this book will be useful as the basis for a course in the geology of the Southwest, especially when supplemented by additional readings from other books and from original literature for in-depth coverage of specific topics. Also, each year many colleges and universities from all over the United States, and in some instances from outside the country, organize field trips to the Southwest, for which this book may serve as a useful introduction.

Finally, because of the tendency in all sciences, and not least in the Earth sciences, toward specialization, it is often difficult to step back and see the broad progress that has been made in understanding a very complex problem. Although many fascinating and important details must be omitted, it is hoped that, through a synthesis of the progress made in studying the geology of the Southwest and in relating the results of relevant studies in other areas and disciplines, the overall paradigm emerges: the geological development of the Southwest.

Acknowledgments

R. E. Bromley, E. Frost, J. M. Hoffer, K. Karlstrom, G. R. Keller, W. R. Muehlberger, S. Semken, and D. Smith all provided important information and valuable assistance in obtaining references and in locating and identifying key outcrops. I thank A. Kron for assistance with Fig. 8.19. A. Downs graciously made the facilities of the Ruth Hall Museum of Paleontology of Ghost Ranch, New Mexico, available and assisted with photography of *Coelophysis* (Fig. 6.8). Several images were provided by the Los Alamos National Laboratory. I thank R. Blakey for permission to use several of his paleogeographic maps. Reviews of an early version of the manuscript by B. Bartleson, R. V. Ingersoll, K. Mabery, and three anonymous reviewers materially improved the content and scope of this book. Detailed reviews of a near-final version by B. Bartleson, R. V. Ingersoll, K. E. Karlstrom, D. Katzman (for Chapter 7 and Box 4.1), and J. Ni (Chapter 8) helped further educate me and significantly improved this manuscript. K. E. Karlstrom shared unpublished manuscripts and information. Finally, I thank my wife Helen Fabel and son Gregory for their patience and good humor in indulging me in the writing of this book. All illustrations were made, and all photographs taken, by the author, unless otherwise noted.

Note on the text

In the title of this book I have used the term 'billion' because common parlance in North America is to speak of the age of Earth in billions of years. To Americans and Canadians the term 'billion' means a quantity equal to 10^9, i.e., a thousand million. However, in large parts of Europe the term 'billion' stands for a million million (10^{12}), a quantity which we in North America call a 'trillion.' Therefore, in the remainder of this book I have avoided the term 'billion' and used instead 'thousand million' or appropriate abbreviations (see List of abbreviations).

All references to direction are with respect to *present* coordinates. Present directions are not necessarily the same (in fact, probably are *not* the same) as when the geologic units under discussion were emplaced. Similarly, references to North America, the Southwest, etc., do not imply that these geographic regions were then as they are today. North America did not achieve anything close to its present outline until after the breakup of Pangea, which began in the Triassic period sometime after 248 million years ago, and has continued to change its shape and area even to the present.

Words or phrases introduced in **bold** font refer to definitions and/or explanations in text boxes.

Abbreviations

m.y. Million years. Used to indicate an interval of time, but not a specific time in the past.

Ma Million (10^6) years ago. From *mega* (million) and *anni* (years). Used for dates of events in the past.

Myr Million (10^6) years or million years old. Used for a duration of time or age.

Ga Thousand million (10^9) years ago. From *giga* (10^9) and *anni* (years). Used for dates of events in the past.

Gyr Thousand million (10^9) years or thousand million years old. Used for a duration of time or age.

kg/m^3 Unit of density in the widely accepted *Système International d' Unités*, or SI, system. Equivalent to $g/cm^3 \times 1000$ in the cgs system.

km Kilometer.

cm Centimeter.

GPa Gigapascal. Unit of pressure equivalent to 10 kilobars.

Introduction

The Southwest evokes images of dusty desert landscapes beset with narrow mountain ranges, of the vast and colorful expanses of the Colorado Plateau and land of the Diné, of Monument Valley, and perhaps of the Spanish and Mexican cultural heritage. Terrains range from barren, seemingly lifeless, deserts to verdant, forested mountains, and vegetation zones from Sonoran to Alpine. Its varied landscapes have challenged explorers and settlers; beckoned artists and adventurers. They may elicit wonderment and awe but can be haunting, even intimidating. The beauty of the Southwest is often stark, typically subtle.

The Southwest of the United States is a region that defies precise geographic definition, eschewing neatly defined physiographic subdivisions. Most usages of the term 'Southwest' include the arid and semiarid region stretching from west Texas across New Mexico and Arizona to southern California (Fig. I.1; see also Plate 1). This vast expanse of desert, the heart of the Southwest, comprises dominantly the Basin and Range and Colorado Plateau physiographic provinces. The Basin and Range province, typifying the southern parts of New Mexico and Arizona and northern Mexico, is that region characterized by a distinctive physiography of narrow mountain ranges separated by broad, sediment-filled desert basins. In contrast, the Plateau is that region of *northern* New Mexico, *northern* Arizona, western Colorado, and much of Utah characterized by broad plateaus, deeply incised canyons, and mainly flat-lying sedimentary strata. In central Arizona, between the low desert basins of Phoenix and Tucson (330–730 m above sea level) and the high plateau of Flagstaff (2100 m) lies a region of ridges and valleys. This area, transitional in its physiography and exposed geology, is referred to by geologists as the 'Transition Zone.' These

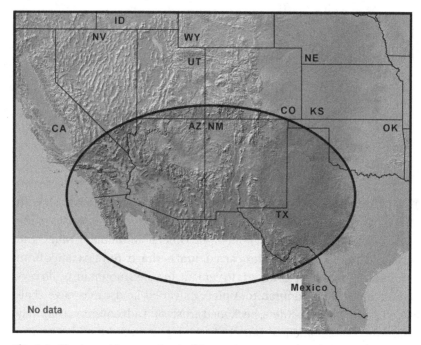

Fig. 1.1. Physiographic map of part of the western USA and northern Mexico compiled from digital elevation data. The Southwest of North America, as loosely defined in this book, is shown in the oval inset. Modified from world-wide web pages by Andrew D. Birrell. See also plate 1.

provinces acquired their physiography only a few tens of millions of years ago, long after most of the crust in the Southwest had formed. The region known as the Southwest includes, as well, part of the southern Rocky Mountains and western Great Plains, the latter being part of the stable core of North America. Arguably, the Southwest includes southern Nevada and Utah, and possibly southern Colorado. Naturally, the Basin and Range and Colorado Plateau provinces constitute a major focus of this book, but relevant parts of adjacent provinces are also included. These tectonic provinces are described further in Chapter 8. Geological features do not stop at international boundaries, and therefore many discussions and examples presented in this book lie in the states of Chihuahua and Sonora of northern Mexico. Typically, however, geological features are better studied on the northern side of the international border. These areas therefore garner more attention in this book.

The Southwest is a land of contrasts. Human presence ranges from log hogans and pre-Columbian adobe villages to state-of-the-art radio telescopes imaging the edges of the universe. Indigenous, Hispanic, and Anglo cultures mingle. Similarly, rocks exposed in the Southwest are diverse and varied. They range from ancient sediment and lava, now recrystallized by burial in the Earth's crust to depths exceeding 20 km, to lava flows so young that they devastated Anasazi cornfields. Volcanism ranged from passive effusion of basalt flows, similar to that occurring in Hawaii today, to catastrophic explosions of silicic magma that devastated wide regions of the continent and probably had global consequences. Sedimentary depositional environments ranged from deep marine to inland sea to, of course, terrestrial. In the Southwest the geology is youthful. Faulting remains active, with its consequent hazard for human existence. Volcanism occurred as recently as one thousand years ago and will happen again. The intersection of geology with social issues is manifest.

Geologic time scale, modified from the Geological Society of America 1998 Geologic Time Scale (A. R. Palmer, compiler). Permian-Triassic and basal Cambrian boundaries are from Bowring and Erwin (1998). Names and ages of subdivisions of the Proterozoic conform to international convention. Only the major time intervals are shown. The shaded region on the time scale that precedes each chapter indicates the time interval discussed in the chapter. Significant geologic events are indicated along the side. The ages of some of the major boundaries remain considerably uncertain, thus undoubtedly the time scale will continue to be modified with the results of more precise age determinations.

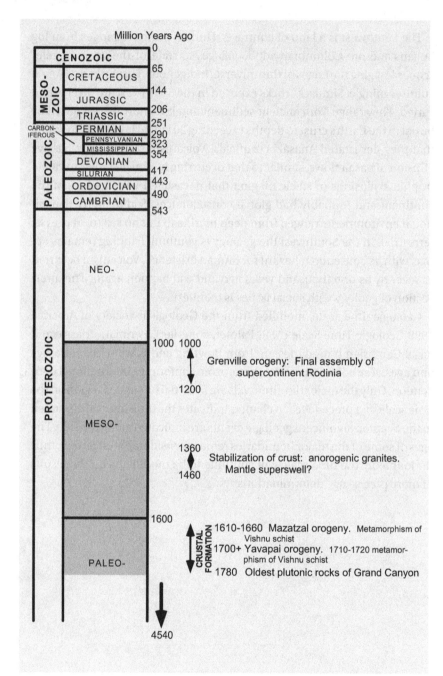

Million Years Ago

CENOZOIC		0
		65
MESO-ZOIC	CRETACEOUS	
		144
	JURASSIC	
		206
	TRIASSIC	251
PALEOZOIC	PERMIAN	290
CARBON-IFEROUS	PENNSYLVANIAN	323
	MISSISSIPPIAN	354
	DEVONIAN	
		417
	SILURIAN	443
	ORDOVICIAN	490
	CAMBRIAN	
		543

PROTEROZOIC

NEO-

MESO-

PALEO-

1000   1000

Grenville orogeny: Final assembly of
supercontinent Rodinia

1200

1360

Stabilization of crust: anorogenic granites.
Mantle superswell?

1460

1600

CRUSTAL FORMATION

1610-1660 Mazatzal orogeny. Metamorphism of
Vishnu schist
1700+ Yavapai orogeny. 1710-1720 metamorphism of Vishnu schist
1780   Oldest plutonic rocks of Grand Canyon

4540

1

Growth of a continent: formation and stabilization of continental lithosphere

Paleoproterozoic to Mesoproterozoic

Cutting through the Gunnison uplift of western Colorado, the Gunnison River exposes rocks of Paleoproterozoic and Mesoproterozoic age in the famous Black Canyon. Seen here are schists and gneiss metamorphosed more than 1.7 Ga. These rocks are among the oldest in the Southwest, forming the basement upon which later, stratified rocks were deposited. Although their exact relationship to rocks exposed in central Arizona is uncertain, they are typical of rocks underlying all of the Southwest but exposed only discontinuously. The name 'Black Canyon' derives from the dark aspect the rocks present.

1.1 Progress in deciphering the Precambrian

The geologic story of the Southwest begins far back in the Precambrian, that period of the Earth's history prior to 543 Ma. It was during the Precambrian that the underpinnings of the crust were formed from the underlying mantle by a variety of processes. Yet events of this great age in the Southwest, as in most other areas of the world, have been difficult to interpret. For more than a century geologists have recognized that Precambrian rocks record events of regional and global importance, but it was not possible until relatively recently to make significant progress in deciphering these events. Earth's Precambrian history remained obscure for several reasons. Over great regions of the continents, rocks of Precambrian age are covered by younger sedimentary rocks, hiding them from direct observation except in deep wells. Where exposed, much of the Precambrian is highly metamorphosed and/or deformed on all scales, destroying or obscuring features of the original rocks that aid in interpreting their genesis. Most importantly, though, rocks of Precambrian age, even where unmetamorphosed, are nearly devoid of the fossils that were essential for the relative age determinations and regional correlations upon which an understanding of geological events was based. Only in the Cambrian period beginning around 543 Ma did organisms suddenly develop the hard body parts that enabled their remains to be preserved well.

Despite the difficulties, tremendous advances have been made in understanding Earth's history before the Cambrian and continue to be made at an increasing rate, not least in the Southwest. Significant progress in understanding the Precambrian awaited the advances in radiometric dating techniques made possible by development of the mass spectrometer. Although the modern mass spectrometer was pioneered in the 1930s, its refinement and commercial availability following World War II led to an explosion of information related to the Precambrian beginning in the 1950s (White and Wood, 1968; Nier, 1989). Several isotopic systems are now routinely used to date crystalline rocks of Precambrian age. Other tools and techniques, such as high-sensitivity electron and ion microprobes and precise $^{40}Ar/^{39}Ar$ dating, allow analysis of individual grains, and even of zones within grains, of minerals such as zircon, xenotime, and monazite. Microanalytical dating and analysis techniques have made it possible to unravel thermal histories and cooling rates of crystalline rocks. In addition, geologists now have a plate-tectonic framework, based on modern analogues, in which to interpret crust-forming events. As a result of these developments, great potential exists for refining our knowledge of the paleogeography and assembly of crustal building blocks, of the process by which newly formed crust is stabilized, and of relationships between plutonism, deformation, and metamorphism of lower and middle crust. Study of older and deeper levels of the crust allow geologists to better understand modern processes.

Although the age of the Earth is now known to be about 4.54 Gyr (Dalrymple, 1991) (Fig. 1.1), the oldest known rocks, from the Northwest Territories, Canada, are 4.055 Gyr old and zircon crystals, from western Australia, are as old as 4.40 Gyr (Froude *et al.*, 1983; Bowring *et al.*, 1989; Wilde *et al.*, 2001). In the United States, the oldest dated rocks are 3.6–3.7 Gyr, although zircons from older rocks incorporated into quartzite are as old as 3.96 Gyr (Mueller *et al.*, 1992). Thus, the term 'Precambrian,' simply assigned to rocks lying stratigraphically below rocks of Cambrian age, incorporates over four thousand million years – almost 88% – of the Earth's history, and the Proterozoic spans almost two thousand million years. Successively younger time units (both eras and periods) tend to represent shorter intervals of time, reflecting simply the greater preservation of younger rocks and the enhanced ability to resolve and interpret more recent events. In the Southwest, rocks are no older than a mere two thousand m.y. (Proterozoic Eon), the age of crustal formation in this region. For this age, the terms 'Precambrian' and 'Proterozoic' are interchangeable (Fig. 1.1).

This chapter, and the next two, focus on Proterozoic rocks of the Southwest. The present chapter describes the formation of continental crust in the Southwest, which occurred in the Paleoproterozoic and Mesoproterozoic. In understanding crustal formation, surface volcanic and tectonic processes must be inferred. Yet most of the Precambrian rocks exposed in the Southwest were formed or metamorphosed, not at or near the surface, but rather in the middle crust, at depths typically of 10–20 km, during assembly of the continent. Thus, a challenge presented to geologists is to interpret surface events from a middle crustal record. Crust-forming events recorded in the Southwest were part of the formation of the larger continent of Laurentia. Laurentia comprised the ancestral **craton** (Box 1.1) of North America, Greenland, parts of Scotland, Scandinavia, and possibly Argentina. For part of the Proterozoic, Laurentia was embedded in a still

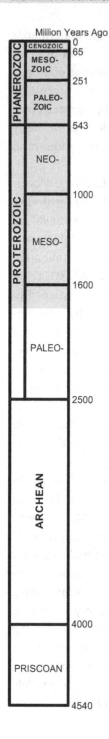

Million Years Ago

Fig. 1.1. In this simplified time scale is shown the period of the Earth's history (shaded) covered in this book. Time before the lowermost unit of the Paleozoic (Cambrian) is typically referred to by the more general term 'Precambrian.' The oldest rocks in the Southwest (approximately 1.84 Gyr old) (Hawkins *et al.*, 1996) are much less than half the age of the Earth.

Box 1.1 Craton The relatively stable interior of a continent, typically composed of Archean and Proterozoic rocks. These ancient rocks are exposed in areas called 'shields' because of their subdued topography and low domal forms, but are elsewhere buried beneath younger sediments and sedimentary rocks. Because of its buoyancy and strength, the craton generally does not experience the rapid subsidence and uplift, and the deformation, which occur at continental margins. Cratons are not wholly without deformation, however. Structural basins, domes, and arches do form on cratons, and major deformational events initiated at margins may extend onto the cratons.

---

larger continent, a *super*continent. It persisted as a continental entity until it was fragmented at the end of the Proterozoic, when the present continents began to take their forms (Dalziel, 1997). Thereafter, the largest part of the original continent of Laurentia retained the name 'Laurentia' until the reassembly of continental fragments into the supercontinent of Pangea.

1.2 Distribution of Proterozoic rocks

Rocks of Proterozoic age *underlie* most of the Southwest. In contrast, they are *exposed* only in limited areas, where uplift creates elevation differences and allows younger rocks to be thinned and stripped away, either by erosion or tectonic processes. Such is the case in many mountain ranges throughout the Southwest. In some places, such as the Grand Canyon of the Colorado River and the Black Canyon of the Gunnison River, a combination of uplift and deep erosion by rivers has exposed Proterozoic rocks.

In the Southwest, Proterozoic rocks are exposed in two major transects that fortuitously strike perpendicular or nearly so to major age boundaries (Fig. 1.2). The first of these is a northwest-trending belt stretching 500 km from Sonora, Mexico, and southeastern Arizona to Nevada and southeastern California. Much of this area lies in the physiographic region referred to as the 'Transition Zone' between the Colorado Plateau in northern Arizona and the Basin and Range province to the south. The second major outcrop belt comprises exposures in the cores of various ranges of the Rocky Mountains extending from southern Wyoming through Colorado and New Mexico to west Texas and northern Chihuahua (Karlstrom and Bowring, 1993). Proterozoic rocks crop out in other areas of the Southwest, of course, but not with the continuity and lateral extent that have allowed major crust-forming events of the Proterozoic to be discerned.

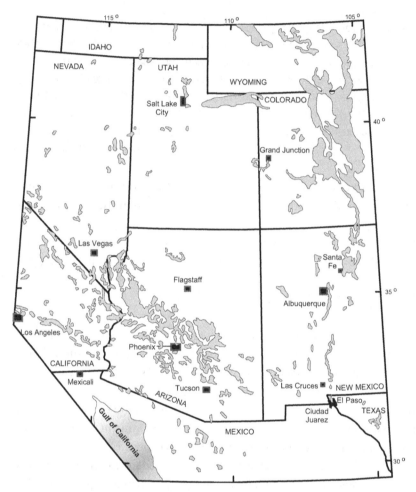

Fig. 1.2. Rocks of Precambrian age underlie younger rocks throughout the Southwest and adjacent regions. They are exposed mainly in two belts, a northwest-trending belt through central Arizona and in ranges of the Rocky Mountains from west Texas to Wyoming. Simplified from Condie (1981).

1.3 Age provinces

To better understand these rocks and to appreciate their significance, it is necessary to relate them to the Precambrian rocks that make up the North American craton as a whole and to understand how the craton was constructed. Much of the present understanding of the processes by which

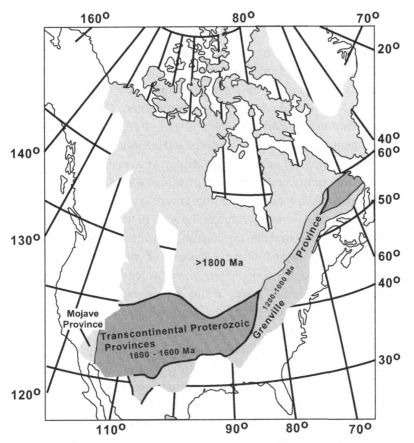

Fig. 1.3. Age provinces of Laurentia. The northern part of the craton (formed before 1.8 Ga) and the Transcontinental Proterozoic provinces (formed 1.8–1.6 Ga) represent crust derived from the mantle at these times. Much of the Grenville province is recycled older material modified 1.2–1.0 Ga. The modern outline of North America is shown for reference, and Greenland has been restored to its position prior to 60 Ma. Large parts of the continent, such as the western parts of the USA and Canada and northern Mexico, were added subsequent to the Proterozoic Eon. Modified from Hoffman (1989) with additions from Ruiz *et al*. (1988).

Proterozoic crust formed throughout the craton is derived from studies in the Southwest, where excellent exposures occur.

The oldest rocks of North America occur in the northern part of the continent, where they are exposed on the Canadian shield and in parts of the northern United States (Fig. 1.3). Ages of rocks comprising the northern part of the craton generally exceed 1.8 Gyr, and large parts are older than 2.5 Gyr (Archean). Southward of the shield is a broad northeast-to east-northeast-trending zone of crust that formed 1.8–1.6 Ga, representing growth of the early 'protocraton' to the southeast. This orogenic zone, called the Transcontinental Proterozoic provinces, is 1300 km wide, from Wyoming to Mexico, and stretches from southeastern California at least to the mid-continent region (Van Schmus *et al.*, 1993). It underlies much of the continental United States. As is discussed in the next several sections, this broad orogenic zone is a composite of many smaller provinces. Formation of the Transcontinental Proterozoic provinces within a period of about 200 m.y. represented a major crust-forming event in the Earth's history, one that is not well understood. Addition of the zone to the protocontinent to the north essentially completed the assembly of the Laurentian continent.

Southeast of the Transcontinental Proterozoic provinces is a zone of rocks 1.2–1.0 Gyr old, the Grenville province, which is highly metamorphosed in many locations. Rocks of this age are best known from the eastern USA and southeastern Canada. Across much of the southeastern and midcontinental USA, rocks of Grenville age are hidden from direct view by much younger rocks. In the Southwest, they are exposed in west Texas and northern Mexico. In contrast with rocks of the Transcontinental Proterozoic provinces, which represent newly formed crust, much of the Grenville province consists of older rocks, which were deformed and metamorphosed during an event referred to as the Grenville orogeny. The amount of new crust added to the craton during the Grenville orogenic event was relatively minor (Mosher, 1993). West of the Transcontinental Proterozoic provinces is the problematic Mojave province, which also seems to represent continental growth by reworking of older material. The Grenville and Mojave provinces are discussed in more detail later in this chapter.

1.4 Tectonostratigraphic terranes

Proterozoic rocks from all areas of the Southwest include a broad range of igneous and sedimentary rocks, although most have undergone at least a moderate degree of metamorphism. Magmatic rocks range from middle-crustal intrusions to volcanic rocks, and compositions from mafic to silicic.

Sedimentary strata include sequences of diverse supracrustal lithologies. Several observations are paramount in understanding the formation of continental crust in the Southwest during the Paleoproterozoic and Mesoproterozoic. First, since the 1960s it has been recognized that rocks from the northwestern part of the Transition Zone in Arizona are consistently older (1.8–1.7 Gyr) than those to the southeast (1.7–1.6 Gyr), although some overlap of ages occurs (Conway and Silver, 1989). Thus were defined two crustal age provinces, known as the Yavapai (the older) and Mazatzal provinces (Karlstrom and Bowring, 1988) (Fig. 1.4).

Second, in recent years it has been recognized that these Paleoproterozoic to Mesoproterozoic rocks comprise a number of **tectonostratigraphic terranes** (Box 1.2), separated from each other by major, generally north- to northeast-trending faults or shear zones (Fig. 1.5). Geologic and geochronologic studies of these terranes yield the conclusion that at least some of these individual terranes are of diverse crustal origin and tectonic evolution and are not necessarily related to adjacent terranes. That is, each block has a slightly different geologic history compared to adjacent blocks. This concept is of key importance in understanding the process by which crust was formed (Karlstrom and Bowring, 1988).

Box 1.2 Terrain, terrane, tectonostratigraphic terrane In contrast to the term 'terrain,' which refers to natural features of land such as topography or vegetation, the term 'terrane' designates a region characterized by a particular rock type or group of related rock types. With respect to the Proterozoic of the Southwest, the term refers to blocks of crust that evolved, at least in part, separately from adjacent blocks and thus have separate crustal origins (Karlstrom and Bowring, 1988). Blocks are bounded by faults and shear zones. Thus, the term 'tectonostratigraphic terrane' emphasizes that the present stratigraphic relationships of the various Proterozoic terranes to each other are tectonically controlled.

Finally, the compositions of many of the magmatic rocks of Paleoproterozoic to Mesoproterozoic age, especially of the hypabyssal intrusions, are calc-alkaline, generally similar to rocks emplaced at modern active plate margins. From the observations taken together, it is possible to infer models for the growth and evolution of Proterozoic crust in the Southwest and to relate them to models for growth of the continent, of which the Southwest is a part, as is discussed in Section 1.5.

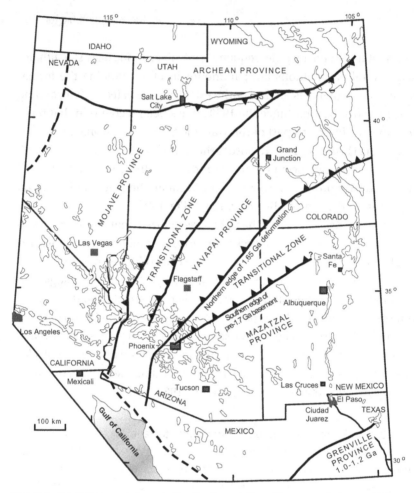

Fig. 1.4. Crustal age provinces in the Southwest. Transition zones between Mojave and Yavapai provinces and Yavapai and Mazatzal provinces represent transitional boundaries, containing slices and slivers of the adjacent provinces. In part, these transitional regions may also result as magmas whose ages are characteristic of the younger province intruded into the pre-existing, older crust, or as younger sediments and volcanic rocks overlapped onto older crust. In addition, in regions where Proterozoic rocks are poorly exposed, the wide transition zones probably reflect lack of information. Barbed lines designate dipping shear zones, with barbs on upper plate. Dashed line is the inferred edge of the Precambrian continent. Light lines enclose outcrops of Precambrian rocks (from Fig. 1.2), which provide control on age boundaries. Notice that province boundaries are extrapolated great distances between the major outcrop belts of central Arizona and the Rocky Mountains. Between areas of outcrop, locations of boundaries are significantly uncertain. From Condie (1981) and Karlstrom (1998, 1999).

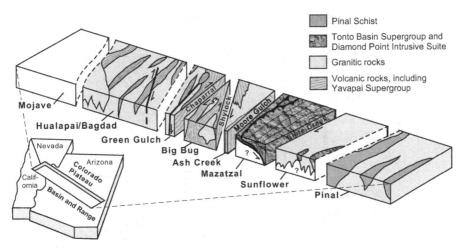

Fig. 1.5. This block diagram shows tectonostratigraphic terranes of the Arizona transition zone. Individual blocks are separated by fault and/or shear zones, some of which are named (between blocks) on this figure. The Moore Gulch shear zone separates blocks of the Yavapai province, deformed and metamorphosed ≥ 1.70 Ga, to the northwest from those of the Mazatzal province, deformed and metamorphosed 1.69–1.63 Ga. Modified from Karlstrom and Bowring (1988).

The following two sections describe these major terranes in some detail because the lithologic character, and in particular the discontinuities in lithologies, are essential for recognizing the terranes and interpreting processes of crustal growth. Although steady progress is being made in understanding the Proterozoic terranes of the Southwest, considerable controversy also exists regarding the paleogeographic setting of different terranes (i.e. whether they represent arc, backarc, or ocean floor) and about how and when terranes came together. Terranes may have been assembled through processes of subduction, collision, transcurrent faulting, or combinations thereof. Thus, the results summarized in these sections are very much a snapshot of a rapidly evolving understanding of Proterozoic crustal growth in the Southwest.

1.4.1 Terranes of Arizona

Proterozoic rocks exposed in the Transition Zone of central Arizona are divided into at least eight separate tectonostratigraphic terranes, or blocks (Fig. 1.5). Rocks of the five northwestern blocks are generally characterized

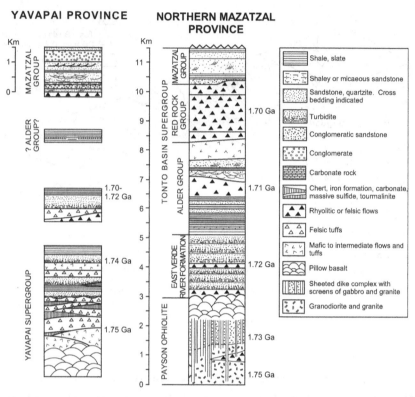

Fig. 1.6. Stratigraphic columns for major units (excluding intrusive suites) of the Yavapai province (left) and northern Mazatzal province (right). The Pinal schist is omitted from the Mazatzal column because its correlation with the Tonto Basin Supergroup is uncertain. Thicknesses shown, where known. From Karlstrom and Bowring (1993).

by a 'basement' of supracrustal rocks 1.76–1.70 Gyr old (although, locally, crust in the northwestern block may be as old as 2.2–1.8 Gyr) and of associated batholiths. Supracrustal basement mainly comprises the Yavapai Supergroup (Fig. 1.6), a diverse sequence of mafic to silicic volcanic and volcaniclastic strata and turbidites, similar to those of modern orogenic zones. Volcanic rocks consist of pillow basalt and andesitic to rhyolitic flows and tuffs. Sedimentary strata include shale, slate, sandstone, and their metamorphosed equivalents. Minor chert and iron formations are also present. Similar supracrustal lithologies in different blocks suggest linkages to each other. The compositions of magmatic rocks intruded into the supracrustal strata are typically mafic to intermediate. The ages of magmatic rocks generally range from 1.79 to 1.66 Gyr. Deformational and metamorphic relations are very complex, and differ in different blocks. The degree of metamorphism ranges from greenschist to amphibolite facies. Although each of these blocks is separate and distinct from adjacent blocks, all have in common the fact that they underwent significant deformation and metamorphism between 1.74 and 1.70 Ga. This common feature is the basis on which the blocks are grouped together into a single 'province,' the Yavapai province (Karlstrom and Bowring, 1988).

Box 1.3 Ophiolite An assemblage of rocks consisting of chert, submarine lava flows, a **sheeted dike complex**, gabbro, and ultramafic rocks (typically deformed and serpentinized) (see Fig. 1.7). The sheeted dike complex consists of a series of parallel dikes, with few or no intervening screens of other rocks, that were successively injected as feeders for the overlying flows. The term 'sheeted' is used to describe the repetition of parallel dikes (geometric sheets) normal to the plane of the dikes; 'complex' refers to the complexity of the intrusive sequence.

Ophiolites represent sections of oceanic crust that have been preserved and exposed to view, typically by having been thrust onto continental margins. In contrast, most oceanic crust is destroyed at subduction zones. Many ophiolites are thought to result from closure of marginal basins, such as backarc or intra-arc basins, therefore they may not be representative of 'normal' oceanic crust formed at mid-ocean ridges. Others, such as that of Macquarie Island located between Australia, New Zealand, and Antarctica (Goscombe and Everard, 1999), however, do seem to be representative of typical oceanic crust, and constitute important models for understanding the structure and composition of oceanic crust, which cannot be studied directly otherwise except by drilling. By comparison to the excellent exposures of essentially the entire crust provided by ophiolites, the deepest drill hole to date has penetrated only to near the expected base of the sheeted dike complex (Detrick *et al.*, 1994).

Supracrustal and magmatic rocks of the terranes southeast of the Moore Gulch fault zone (Fig. 1.5) are consistently somewhat younger, predominantly ranging in age from 1.71 to 1.69 Gyr. Thus, the Moore Gulch fault zone coincides with a major crustal age boundary. This southern province consists of supracrustal strata, such as graywacke, slate, tuff, and rhyolitic and basaltic flows, and of associated hypabyssal intrusions. Metamorphism and deformation in these blocks are inferred to result from a major tectonic event called the Mazatzal orogeny, 1.66–1.60 Gyr old (Karl-strom and Bowring, 1988). The stratigraphy of the Mazatzal province is better known than that of the Yavapai province because the rocks are less deformed and have undergone a lower grade of metamorphism (Figs. 1.5 and 1.6). Within the Mazatzal province, the Mazatzal block is recognized as part of the foreland and foreland thrust belt of the Mazatzal orogen. The Sunflower block, characterized by more ductile, higher-grade metamor-phism, and by plutonism, represents the deeper roots of the orogen.

The Mazatzal block has been extensively studied and is illustrative of rock types of other blocks within the Mazatzal province. Overall, the Mazatzal block appears to represent a magmatic arc 1.73 Gyr old that underwent extension during a series of tectonic events (1.75–1.63 Ga), including for-mation of an **ophiolite** complex (Box 1.3), development of a thick sed-imentary basin, and deformation associated with a foreland thrust belt (Karlstrom and Bowring, 1993). The basement of the arc comprises granitic rocks 1.75–1.76 Gyr old (presumably subarc batholiths), occurring partly as pendants in the younger ophiolite. The basement was intruded 1.74–1.73 Ga by gabbroic, dioritic, and tonalitic magmas (Fig. 1.8). In places, these plutonic rocks are overlain by a **sheeted dike complex** (see Box 1.3) (Figs. 1.9 and 1.10), which in turn is overlain by submarine mafic volcanic rocks. Identification of these latter is based on recognition of pillow basalt. The compositions of the 1.74–1.73 Gyr-old plutonic rocks, together with the sheeted dike complex and overlying submarine volcanic rocks, are in-terpreted as an ophiolite complex, a segment of oceanic crust that may have been formed in an intra-arc basin along an arc-parallel strike-slip fault (Dann, 1991, 1997). No ultramafic (mantle) rocks are exposed. Strata overlying the ophiolite comprise a thick sequence of graywacke and silt-stone of turbidite origin (the East Verde River Formation) deposited about 1.72 Ga (Fig. 1.6). In turn, these rocks are overlain by intercalated silici-clastic, volcaniclastic, and volcanic (basalt, andesite, and rhyolite) rocks (Alder Group). These strata were deposited in the intra-arc basin prior to collision and exposure of the ophiolite during the Yavapai orogeny (Dann, 1997).

These lower units of the Mazatzal block are overlain by a sequence 2 km thick that includes ash-flow tuffs and flows of rhyolitic composition (Red Rock Group). The rhyolitic ash-flow tuffs, in particular, are inferred to have been erupted from multiple calderas. The lower units are separated

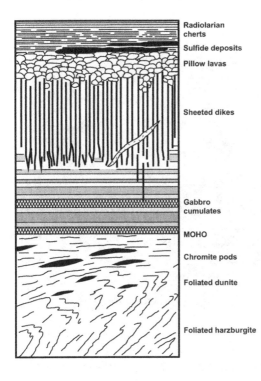

Radiolarian
cherts

Sulfide deposits

Pillow lavas

Sheeted dikes

Gabbro
cumulates

MOHO

Chromite pods

Foliated dunite

Foliated harzburgite

Fig. 1.7. This simplified figure shows the major components of an ophiolite, interpreted as a section through the oceanic crust into the uppermost mantle (modified from Philpotts, 1990).

from overlying shale and quartzite of the Mazatzal Group by a 1.71–1.70 Gyr-old unconformity. These upper sedimentary rocks possibly represent a synorogenic sequence deposited in response to the Yavapai orogeny to the northwest. Finally, all of these magmatic and sedimentary rocks were deformed and metamorphosed in a foreland thrust belt *c.* 1.69–1.65 Ga, with structure dominated by northwest-directed thrusts.

The Paleoproterozoic history of the upper Granite Gorge in the Grand Canyon is similar to that of the Yavapai province in central Arizona, but in the Grand Canyon rocks preserve evidence for a pre-1.8 Gyr-old crust (Hawkins *et al.*, 1996). First, the Elves Chasm pluton, dated at 1.84 Gyr, exhibits both intrusive and gradational contacts with surrounding country rocks, therefore the extent of pre-1.84 Gyr old crust is uncertain. Second, grains of zircon with minimum ages of 2.178–1.998 Gyr are contained within the Tuna Creek granodiorite. Both intrusions provide evidence for the presence, at least locally, of pre-1.84 Gyr old crust. The extent to which such crust underlies the Colorado Plateau is unknown.

Fig. 1.8. These bouldery outcrops of a 1.738 Gyr-old diorite near Payson, Arizona, are part of the Gibson Creek batholith, probably formed in a magma chamber that once underlay an island arc volcano (Conway *et al*., 1987; Karlstrom and Bowring, 1988), much as found today in Japan or other Pacific island chains. The Gibson Creek batholith is part of the Mazatzal terrane (see text). Inset: the minerals comprising the Gibson Creek batholith consist mainly of plagioclase (light mineral) and amphibole (dark mineral).

1.4.2 Terranes of New Mexico and southern Colorado

Paleoproterozoic to Mesoproterozoic rocks of New Mexico and southern Colorado, every bit as complex as those of Arizona, are characterized by a great range of rock types and by superposed, overlapping deformational events. These terranes are particularly difficult to define and to relate to each other because of their discontinuous exposure in isolated mountain ranges. Although widespread agreement exists among different investigators concerning some characteristics of these terranes, interpretations differ regarding how many separate tectonostratigraphic terranes are represented and the relationships among them (Grambling *et al*., 1988; Daniel *et al*., 1995). Several of the major terranes are described here, but not the details of the competing interpretations, which are likely to change anyway.

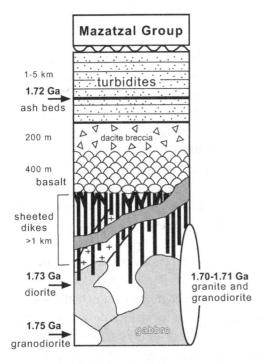

Mazatzal Group

1-5 km
1.72 Ga
ash beds

turbidites

200 m

dacite breccia

400 m
basalt

sheeted
dikes
>1 km

1.73 Ga
diorite

1.70-1.71 Ga
granite and
granodiorite

1.75 Ga
granodiorite

gabbro

Fig. 1.9. Schematic stratigraphic section of the northern Mazatzal Mountains, showing the Payson ophiolite and overlying Proterozoic units. The Payson ophiolite is interpreted as having formed in an intra-arc basin along a strike-slip fault zone parallel to a volcanic arc (Dann, 1997). Modified from Dann (1991, 1997).

Fig. 1.10. Several vertical dikes, < 0.5–1 m in width, are seen in this photograph along Rock Creek in the Mazatzal Mountains of Arizona. The mafic dikes, interpreted to represent oceanic crustal layer 2B (see text), are part of the sheeted dike complex of the Payson ophiolite (Dann, 1991, 1997). The ophiolite represents a segment of oceanic crust formed in a back-arc or intra-arc basin and thrust upward to higher levels in the crust during the Mazatzal orogeny. The boundaries of several dikes are indicated by short vertical white lines.

Considerably more study will be required before a satisfactory understanding of the Proterozoic terranes of New Mexico and Colorado and their relationship to terranes of the Arizona Transition Zone are realized.

The oldest rocks in New Mexico, incorporating the Pecos 'greenstone belt' and other terranes, are 1.76–1.72 Gyr-old mafic metavolcanic sequences that include amphibolite, mafic schist, metarhyolite, calcsilicates, minor banded iron formation, and massive-sulfide deposits. These terranes are intruded by granodiorite plutons 1.75–1.72 Gyr old. Locally, amphibolite preserves pillow structures. These rocks are interpreted to have formed in an oceanic arc or back-arc setting (Soegaard and Eriksson, 1985).

Overlying this mafic metavolcanic basement is a sequence of mainly felsic metavolcanic and metasedimentary strata (Vadito and Hondo Groups, together comprising the Truchas terrane) recognized in most of the uplifts of northern New Mexico (Fig. 1.11). These rocks include a distinctive, kilometer-thick quartzite unit, the 1.70 Ga Ortega Formation, which accumulated on a shallow, south- to southeast-sloping marine shelf. Cross beds, reactivation surfaces, ripple marks and current indicators, and other features record wave, tidal, and storm processes (Fig. 1.12). Parts of the North and Bering Seas may be modern counterparts for the setting in which the Ortega Formation was deposited (Soegaard and Eriksson, 1985). The Ortega Formation probably correlates lithologically and stratigraphically with similar quartzite of the Uncompahgre Formation in the Needle Mountains of southern Colorado (Van Schmus *et al.*, 1993).

Proterozoic rocks of central and southern New Mexico are characterized by younger (1.67–1.60 Gyr old) felsic to mafic metavolcanic schist, quartzite, granitic plutons, and minor pelitic schists (Santa Fe and Manzano terranes). The calc-alkaline composition of plutonic rocks suggests that they may have originated in an island arc.

The relationship of Proterozoic terranes in New Mexico and southern Colorado to terranes exposed in the Transition Zone of Arizona is not at all clear. The general northeasterly strike of Proterozoic faults and shear zones separating terranes in Arizona, as well as the continuity of northeast-striking geophysical anomalies, suggest that the Yavapai and Mazatzal terranes of Arizona are correlative along strike with similar rocks of New Mexico and Colorado. Possibly the 1.67–1.60 Gyr-old rocks of the Santa Fe and Manzano terranes may correlate with the Mazatzal crustal province,

SE   Ground surface (schematic)   **NW**

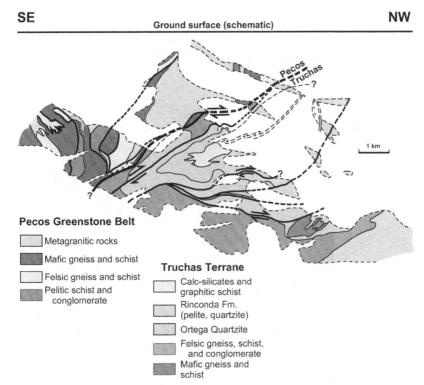

Pecos Greenstone Belt

- Metagranitic rocks
- Mafic gneiss and schist
- Felsic gneiss and schist
- Pelitic schist and conglomerate

Truchas Terrane

- Calc-silicates and graphitic schist
- Rinconda Fm. (pelite, quartzite)
- Ortega Quartzite
- Felsic gneiss, schist, and conglomerate
- Mafic gneiss and schist

Fig. 1.11. Proterozoic rocks exposed in mountain ranges in northern New Mexico are intensely deformed, as shown in this cross section of the Rio Mora and part of the Pecos uplift. Rocks of the older Pecos terrane have been thrust over those of the Truchas terrane, with a minimum displacement of 10 km. Simplified from Grambling *et al.* (1988).

whereas older rocks to the north may be equivalent to the Yavapai crustal province (Daniel *et al.*, 1995).

1.4.3 Mojave province

West of the terranes of Arizona and New Mexico is a region of Paleoproterozoic strata referred to as the Mojave province, exposed across the Mojave Desert and Basin and Range province of southern California, southern Nevada, and central Utah (Fig. 1.4). Rocks of the Mojave province are distinguished from adjacent Paleoproterozoic rocks primarily by differences in Pb and Nd isotopes, which yield generally older ages. This region is less well understood than the terranes described above, for several reasons.

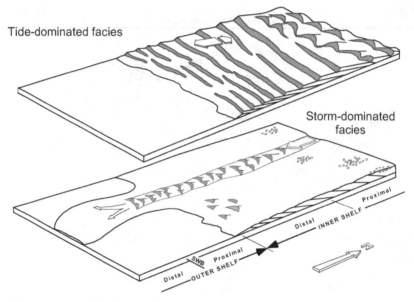

Tide-dominated facies

Storm-dominated facies

Fig. 1.12. Depositional model for quartzite of the Ortega Formation in northern New Mexico. The quartzite is inferred to represent sandstone deposited on a gently sloping, shallow marine shelf, as shown in these figures (deeper water is to the left). The upper figure shows features developed during fair-weather, tide-dominated conditions. Tide-generated, large-scale dunes are present in shallow water. In deeper water, sand waves and storm-generated cross-stratification are subdued by tidal currents. The double-headed arrow suggests reworking by waves and tides of offshore-transported sand. The lower figure shows the spatial relationship between storm-produced facies. On the proximal inner shelf, storms leave behind a lag of conglomerate (irregular stipple). Currents scour broad, shallow channels on the inner shelf while transporting sand onto more distal reaches of the shelf (arrows). Mound-like features on the distal inner shelf represent hummocks which are largely reworked within a few days by tidal processes. SWB, storm-wave base. Modified from Soegaard and Eriksson (1985).

First, outcrops are generally less extensive and therefore less continuous than in the Transition Zone of Arizona and the Rocky Mountains. Protero-zoic rocks are exposed mainly in the cores of mountain ranges through-out the desert of the Mojave and the Basin and Range province. Lack of continuous exposures makes it difficult to map out larger-scale structural features. Second, many of the outcrops are out of place, occurring in the upper plates of Mesozoic thrust faults with tens to possibly hundreds of kilometers of northeastward transport. Finally, many of the rocks are highly metamorphosed and structurally deformed, which destroyed their original lithologic character and field relationships (Wooden and Miller, 1990; Rämö and Calzia, 1998). The generally poorer exposure of Paleoproterozoic rocks throughout this region has made it difficult to recognize tectono-stratigraphic terranes.

None the less, the general evolution of at least the southern part of the Mojave province is understood. As in other parts of the Southwest, the Paleoproterozoic rocks consist of crystalline rocks: gneiss, schist, and migmatite. Supracrustal rocks of the Mojave province were intruded by granodioritic to granitic magmas over a period of time exceeding 100 Myr (1.76–1.64 Ga). Midway through this episode, approximately 1.71–1.70 Ga, the intrusive rocks and their enclosing sedimentary and volcanic rocks underwent orogeny, called the Ivanpah orogeny, characterized by strong deformation and high-temperature, low-pressure metamorphism (Wooden and Miller, 1990). This orogenic event transformed the rocks into complex assemblages of gneiss and migmatite. At least the southeastern part of the Mojave province, and perhaps the entire province, is distinct from the adjacent Yavapai province in that supracrustal rocks, other than quartzites, contain grains of zircon, which preserve ages of up to 2.3 Gyr. In addition, isotopes of Nd are interpreted to indicate that a component of the magmatic rocks is as old as 2.6 Gyr, making up as much as 30–40% of some rocks (Rämö and Calzia, 1998). Most likely this Archean component was introduced as sedimentary detritus, which was subducted and mixed with juvenile material (i.e. material newly extracted from the asthenosphere) at a convergent zone (discussed in the next section). The Mojave province may possibly represent a 'microcraton' rafted in from somewhere distant, or it may have resulted from direct crustal growth in its present position relative to adjacent provinces of the Southwest and the Rocky Mountains (Wooden and Miller, 1990; Rämö and Calzia, 1998).

Just when and how the Mojave and Arizona crustal provinces were juxtaposed is unknown at present. Once in proximity, however, the Mojave province may have been sutured to the Yavapai during regional deformation between 1.74 and 1.72 Ga. Both provinces may then have been accreted to Proterozoic terranes farther north during the Yavapai orogeny 1.72–1.68 Ga (Duebendorfer *et al.*, 2001).

1.5 Accreted terranes at convergent margins

To this point, no overall interpretation of the Proterozoic of the Southwest has been made, although reference to island arcs and back-arc basins, emplacement of ophiolites, etc., has been used in discussion of separate blocks and terranes. Interestingly, the overall paradigm by which to understand crustal formation in the Southwest emerged from study of Mesozoic and Cenozoic rocks of the western USA (described in Chapters 4–7), and was only later applied to the Proterozoic. Geologists recognized, primarily in the 1970s, that much of the western USA consisted of 'suspect' or 'exotic' terranes (various terms were used), consisting of blocks of rocks that did not form in contiguity with adjacent rocks. Rather, they formed in some distant location and were moved to their present location, 'assembled' together, and 'accreted' to the continent by tectonic processes (Burchfiel and Davis, 1972, 1975; Dickinson and Snyder, 1978; Dickinson, 1981; Karlstrom and Bowring, 1988; Bowring and Karlstrom, 1990). But how are blocks of crust formed, and how are they moved from one location to another?

The process by which blocks and terranes were assembled was inferred to be that of subduction at convergent zones. Subduction results in formation of chains of volcanoes formed on the margin of the overriding plate. If the overriding plate consists of oceanic lithosphere and is, hence, submerged, the volcanoes form a chain or arc of islands: an island arc. If the overriding plate is continental, then generally the volcanoes form a lofty range of volcanic mountains, such as the modern Andes of South America. The magma that rises to produce the volcanoes originates from two sources. The main source is from partial melting of the overriding mantle (the 'mantle wedge'), which is fluxed by fluids expelled from the subducting slab. A secondary but none the less significant source of magma may be the subducted oceanic crust itself, which can also partly melt if sufficiently hot (Sigmarsson *et al.*, 1998). In either case, island arcs and associated basinal sedimentary strata, carried on the backs of subducting slabs, can get shoved against continental margins and scraped off as the slab descends into the mantle (Fig. 1.13). Each arc has a characteristic lithology or set of lithologies, and hence constitutes a terrane. In some cases, mid-oceanic islands, oceanic plateaus, or even slivers of continental crust from far distant regions may also be carried by subduction to collide with continental margins.

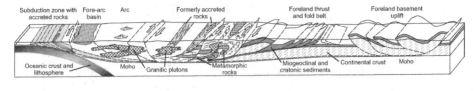

Fig. 1.13. The complexity of a convergent margin is indicated in this conceptual diagram (upper panel), showing accretion of ocean floor crustal rocks to a continental margin. Tectonic elements include: (1) an accretionary 'wedge,' whereby deep-sea sediments and oceanic crust are plastered to the continental margin; (2) a volcanic arc; (3) formerly accreted rocks emplaced at earlier positions of the subduction zone; (4) a metamorphic zone of reworked crust; and (5) a thrust and folded zone, including basement uplifts. The lower panel highlights the major decoupling zones, stripped of the rock units involved in the deformation. From a deep decoupling zone at the base of the crust emanates a series of detachment zones and faults. The diagram was originally drawn to represent the Cordillera of western North America and includes a component of lateral slip parallel to the margin, but it may equally well represent processes by which the continental crust of the Southwest was created in the Proterozoic. Modified from Oldow *et al.* (1989).

Modern analogues for the process of collision and accretion are found at several locations around the rim of the Pacific Ocean. South of Taiwan, oceanic crust of the South China Sea is being subducted eastward beneath the Philippine Sea plate along the Manila trench. Subduction brings the Luzon volcanic arc, a chain of volcanoes stretching 1200 km northward from the Philippines to Taiwan, into contact with the continental shelf of mainland China, on the Eurasian plate. The mountains of Taiwan are a result of the arc–continent collision. In the process, the Luzon arc is also being deformed by thrust- and strike-slip faulting and by vertical-axis block rotation. Eventually, the arc will be 'accreted' to the Chinese continental crust (Lallemand and Tsien, 1997).

A tectonically much more complex, but more realistic, analogue for the process of collision and accretion in the Southwest is found in the southwestern Pacific Ocean 2500 km south of Taiwan. Here, numerous

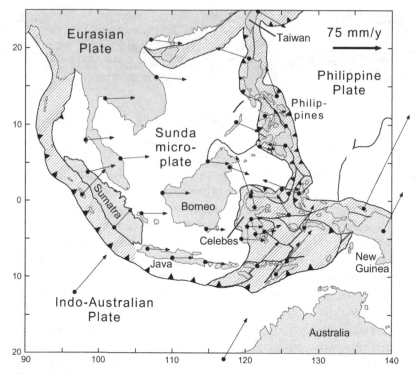

Fig. 1.14. The complex relations of the various plates of the southwestern Pacific Ocean is illustrated in this tectonic map. The Sunda microplate, surrounded on three sides by subduction zones, is being thrust against the Indo-Australian plate. Arrows indicate the directions of movement of the various plates. Lines indicate major tectonic boundaries, including (barbed lines) subduction zones. The diagonally ruled area is a zone of crustal deformation. Modified from Wilson *et al.* (1998).

volcanic islands and island arcs, such as Sumatra, Borneo, the Philippine Islands, and the Solomon Islands, lie above 'stacked' subduction zones scattered across the Philippine plate and the Sunda microplate (Fig. 1.14). Together with their intervening basins, filled with volcanic strata and volcaniclastic and pelagic sediments, these volcanic islands are being swept together (i.e. are colliding with each other 'at sea') as the plates on which they ride carry them inexorably toward subduction zones. Because the volcanic islands and island chains are composed of rocks less dense than oceanic crust, they are not subducted.

The continent of Australia, located solidly on the Indo-Australian plate, is beginning to play a major role in crustal accretion in this complex region. From measurements of absolute plate motion using the satellite-based Global Positioning System, it is determined that the Indo-Australian plate is moving northward at the geologically rapid rate of some 10 cm per year. It is thus colliding rapidly and obliquely with the eastward-moving Sunda microplate, which is being deformed and broken up over a wide zone adjacent to its southeastern margin. Already, some of the islands on the southern edge of the Sunda plate, including Flores, Timor, western New Guinea, and others in between, are currently being accreted to the Indo-Australian plate (Wilson *et al.*, 1998). Other islands or island arcs, such as Java, Borneo, and Celebes, may be added as Australia presses northward. New Britain, the Solomon Islands, and other nearby islands may eventually collide with New Guinea (Phinney *et al.*, 1999). In many cases, these islands may collide with each other in a series of small orogenies before encountering and being accreted to continental crust.

This concept of accretion of island arcs and adjacent basins may, then, explain many features of Paleoproterozoic to Mesoproterozoic rocks of the Southwest (Fig. 1.15; see also Plate 2), such as the generally calc-alkaline compositions of magmatic rocks, their close spatial and temporal association with volcaniclastic strata, and tectonically incorporated sections of ocean floor, such as the Payson ophiolite (Dann, 1991, 1997). The model may also explain the difference between age of *formation* of the rocks, and age of *deformation*. In many cases, adjacent blocks have similar ages and were deformed together prior to a succeeding major deformational event, as though discrete oceanic islands were swept together prior to colliding with continental crust. In addition, magmatic arcs may form on older collisional orogens, creating overlaps in ages. By this means, continental crust may have been assembled during several episodes of convergent tectonism. An important conclusion to be drawn from this mechanism is that plate-tectonic processes as we understand them today were operating vigorously during the Paleoproterozoic and Mesoproterozoic. Many features of the present lithosphere in the Southwest, such as the orientations of major fault and fracture zones and patterns of deformation, crustal thinning, and magmatism, may reflect an 'inheritance' from the time of lithospheric formation (Karlstrom and Humphreys, 1998).

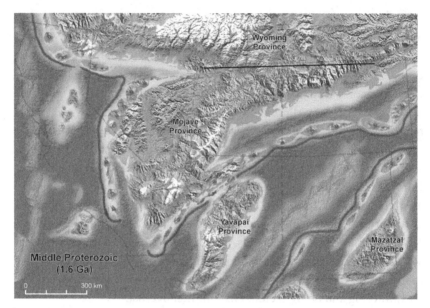

Fig. 1.15. Using the tectonics of the southwestern Pacific Ocean (Fig. 1.14) as an analogue, this imaginative paleogeographic map illustrates how Proterozoic crust in the Southwest may have been created. An array of oceanic islands and island arcs are swept together as they are carried toward their respective subduction zones adjacent to the continental margin. Eventually they collide with, and are accreted to, the continent. This figure illustrates a stage in the development of Laurentia just prior to formation of the Mesoproterozoic Yavapai and Mazatzal terranes of the Southwest. From R. C. Blakey, website: http://vishnu.glg.nau.edu/rcb/paleogeogwus.htm, by permission. See also Plate 2.

1.6 Stabilization of crust

Beginning about 1.6 Ga, after some 200 Myr of accretion of island arc-terranes and related basinal strata along a convergent continental margin, the Proterozoic craton apparently entered an era of relative tectonic and magmatic quiescence throughout the Southwest. This hiatus, which lasted at least 100 Myr, is interpreted to indicate that during this time an important change in the style of tectonism occurred (Karlstrom *et al.*, 2001). However, considerable uncertainty exists regarding tectonic events in the subsequent period, between 1.5 Ga and the Grenville orogeny, 1.2–1.0 Ga (next section). During this 300 Myr period of time, the Proterozoic crust in

Fig. 1.16. The 1.44 Gyr-old Sandia Granite, here shown cropping out as rounded boulders on the western flank of the Sandia Mountains near Albuquerque, is representative of the huge volumes of magmatic rocks intruded into the middle crust after major crustal formation in the Southwest was complete. Inset photo shows phenocrysts of potassium feldspar, up to 3.5 cm in size, and clots of mafic rock. The clots probably represent basaltic melts that were intruded into the granitic magma chamber, in some cases mixing with the granitic magma. Pen is 14.5 cm long.

the Southwest, newly assembled in the Yavapai and Mazatzal orogenies, continued to undergo considerable modification, involving deep-seated melting with subsequent intrusion of granitic magmas into higher levels of the crust. Although several pulses of magmatism are known, intrusion occurred dominantly 1.46–1.36 Ga, and most of that between 1.45 and 1.40 Ga (Fig. 1.16). Generally, the magmas, originating in the lower crust, lodged in the overlying crust at depths of from 8 to 17 km. Magmatism spanned much of the continent, from California at least as far east as Illinois and Wisconsin, and intrusions as far east as Quebec and Labrador may be related (Fig. 1.17). Magmatism constituted such a profound thermal event that between 15 and 40% of exposed upper Proterozoic crust formed at this time (Anderson, 1989).

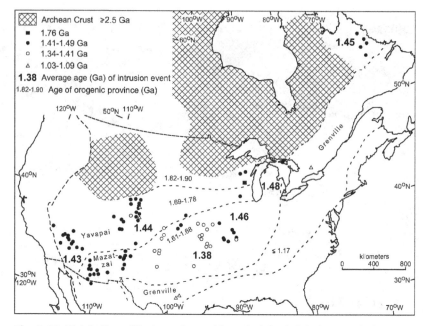

Fig. 1.17. Distribution of Proterozoic granitic rocks intruded during a period of crustal stabilization following major crustal accretion of the Yavapai and Mazatzal orogenies. In the Southwest, intrusion mainly occurred 1.46–1.36 Ga. Modified from Anderson (1989).

It is probably safe to infer that the impressive scale of the magmatic event was unique because the preceding crust-forming event, in which a zone of crust 1300 km wide was formed in the comparatively short span of 200 Myr, was also unique. Rocks forming the crust, assembled in the preceding Yavapai and Mazatzal orogenies, probably contained a large proportion of biotite, amphibole, and other water-bearing minerals with lower melting temperatures than their anhydrous equivalents. Hence, the crust was primed for a magma-generating event. Heating of the crust from below resulted in melting of more fusible components and their transfer to the middle and upper crust. If continents grow by accretion of intra-oceanic arcs, as generally thought, then magmatic events such as recorded during this period of time may be required to form crust that is typical of cratons today. Modern island arcs have bulk compositions that are generally basaltic, i.e. that are very different from the generally andesitic composition that characterizes mature continental crust (Holbrook *et al.*, 1999). In order to transform generally mafic, island-arc-type crust into more typical continental crust, the middle and upper crust must become substantially more silicic, and much of the mafic lower crust must be removed. This process may occur by remelting of mafic lower crust, augmented by fractional crystallization of the derived melts, to form andesitic to rhyolitic melts, which then rise buoyantly to higher levels in the crust. Heating and remelting of lower crust may be driven by 'underplating' of mantle-derived mafic melts to the base of the crust. Alternatively, mafic and/or ultramafic lower crust may be removed by delamination (Holbrook *et al.*, 1999). In either case, the result of the profound, 1.46–1.36 Gyr-old magmatic event was probably to move thermally unstable lower crustal components (i.e. hydrous and incompatible-element-enriched) to the upper crust and to remove or reform mafic and ultramafic lower crust. Thus, the overall effect was a more mature and stable continental crust, typical of cratons.

Because the intrusions occurred within the continent far from inferred orogenic margins, and because of geochemical evidence for derivation from mainly crustal sources and general absence of coeval tectonism, they have been considered to be anorogenic in their association. 'Anorogenic' means that the intrusions lacked any relationship to orogenic deformation, and that the main strain affecting the upper crust was simply that required to accommodate the rise of the magma bodies. The origin of the magmatic belt has been variously attributed to incipient intracontinental rifting or to an enormous mantle plume. If this anorogenic model is correct, then it implies an event that is probably unique in the Earth's history. No analogue for this sort of event is recognized from Phanerozoic time.

However, recent detailed studies of the granitic intrusions and their margins, and particularly of associated mylonitic shear zones, have yielded evidence that emplacement of the intrusive bodies occurred in a regional compressive or right-lateral transpressive tectonic setting (Nyman *et al.*, 1994; Karlstrom *et al.*, 2001). Thus, the magmatism may be fundamentally 'synorogenic' rather than 'anorogenic.' But what kind of tectonic event could give rise to magmatism up to 1300 km inboard of the present plate margin, and over a relatively long period of time? Based partly on this new work, on inferences from the magmatic and tectonic record of better-exposed areas in Scandinavia, Australia, and eastern Canada, and on analogy with back-arc processes behind the Cordillera of North and South America, a new paradigm is emerging that is more compatible with the preceding collisional events of the Yavapai and Mazatzal orogenies and with the subsequent (next section) Grenville deformation (Nyman *et al.*, 1994; Duebendorfer and Christensen, 1995; Rivers and Corrigan, 2000).

It is proposed (Rivers and Corrigan, 2000; Karlstrom *et al.*, 2001) that convergence along the southern plate margin was more or less continuous from the Mazatzal to the Grenville orogeny, i.e. that the southern margin of Laurentia was a long-lived (from *c.* 1.8 to 1.0 Ga) convergent margin characterized by a series of magmatic arcs. The convergent zone stretched from New Mexico and west Texas at least to Labrador, a distance of almost 4000 km. Orogenesis involved both continental-arc-magmatism and postulated collision of an approximately 1.50–1.40 Gyr-old juvenile crustal block. These rocks were subsequently overprinted by the Grenville orogeny and concealed beneath Paleozoic strata. The fact that magmatism was extensive over a broad region and over a long time interval suggests that, in detail, a variety of tectonic settings could have existed, including transcurrent faulting and multiple regions of back-arc extension. However, overall the setting was subduction-related.

Moreover, this *c.* 1.8–1.0 Gyr-old zone of continental accretion may have been part of an even longer collisional boundary, stretching northeastward from Laurentia to Scandinavia and westward along the Australian craton. This reconstruction is discussed in detail in Chapter 3. If so, then the proposed southern margin of Laurentia would have been greater than 10,000 km in length, similar in length to the modern Andean arc (Rivers and Corrigan, 2000) and still much smaller than the modern Cordilleran convergent system of North and South America. Needless to say, this tectonic model for events in the interval between *c.* 1.6 and 1.2 Ga will continue to be developed and evaluated with further geochronologic and other detailed studies.

1.7 Final assembly: the Grenville orogeny

The final period of Proterozoic crustal formation, represented by the Grenville province (Fig. 1.3), followed the major crust-forming orogens of the Yavapai and Mazatzal events. The Grenville province is generally thought to record more than 300 Myr of mountain-building activity, culminating in the collision of one or several arcs and then of an entire continent with the southern and southeastern margin of Laurentia 1.15–1.12 Ga (Soegaard and Callahan, 1994; Mosher, 1998). In the Southwest, rocks of Grenville age are exposed in central and west Texas and in northern Chihuahua (Ruiz *et al.*, 1988) (Fig. 1.4). They are presumably continuous in the subsurface from the Southwest through the Appalachian Mountains to the maritime provinces of southeastern Canada. From central Chihuahua southward, rocks of Proterozoic age are not part of the Laurentian craton. Rather, they are younger (late Neoproterozoic) (Lopez *et al.*, 2001) and are probably exotic, having been juxtaposed against the Laurentian craton during the formation of Pangea (Chapter 5).

In Texas, Grenville-age rocks are exposed in the Franklin Mountains near El Paso, in the Carrizo Mountains and adjoining areas near Van Horn, and in the Llano uplift of central Texas. These exposures, considered together, are interpreted in two different but related ways by different investigators. The first interpretation considers that the exposures are part of a single transect across the Grenville orogen from a core in the Llano area to the northern margin near Van Horn to the undeformed southern margin of Laurentia near El Paso (Mosher, 1998). The second interpretation concurs that Grenville rocks of the Llano area represent the core of a collisional orogen, but relates Grenville rocks of the Van Horn and El Paso areas and others as far northwestward as Death Valley to a zone of transcurrent faulting and rifting normal to the Grenville deformational belt (Bickford *et al.*, 2000). Resolving these two contrasting ideas will require additional and detailed studies.

No matter which of the above interpretations is correct, the structure of the Grenville orogen is generally that of intensely deformed thrust sheets, which transpose highly metamorphosed volcanic and volcaniclastic strata northward over a sequence of pre-orogenic, shallow-marine strata of the continental margin. Undeformed and relatively unmetamorphosed strata, approximately 1.26 Gyr old, are exposed in the Franklin Mountains, near El Paso, in the foreland of the orogen, which lay inboard of and relatively far from the Grenville front. These strata are described in Chapter 2.

In the Van Horn region of Texas, both the tectonic foreland and the hinterland of the Grenville orogen are preserved. Within a distance of less than 7 km, the entire transition zone from deformed allochthonous thrust sheets to undeformed autochthonous strata occurs. A shallow-marine sequence of sedimentary and volcanic strata, correlative to that of the Franklin Mountains, is exposed beneath a series of thrust sheets, but here they are deformed into tight folds. The main thrust sheet (Streeruwitz thrust), dipping southward, places the shelf sedimentary strata beneath a sequence of highly metamorphosed sedimentary strata, rhyolite flows, and welded ash-flow tuffs, possibly of the autochthonous block of a southern, impacting continental block (Mosher, 1998) or alternatively of the southern margin of Laurentia (Bickford *et al.*, 2000). A synorogenic clastic wedge (Hazel Formation) of terrestrial sediments was deposited to the north in a sedimentary basin adjacent to the developing orogen. The deformation that created the highland source and the basin in which the Hazel Formation accumulated occurred 1.10–1.08 Ga. Movement on the Streeruwitz thrust, which post-dates deposition of the Hazel Formation, occurred about 1.035 Ga (Bickford *et al.*, 2000). Together, these structures comprise a narrow fold-and-thrust belt and foreland basin, which are interpreted to have formed near a convergent margin (Soegaard and Callahan, 1994; Mosher, 1998).

The most extensive exposures of the Grenville orogen occur in the Llano uplift (popularly known as the 'Hill Country') of central Texas, 600 km east of Van Horn. The uplift, with an area of about 3800 km², exposes a collisional suture between an exotic arc terrane and continental crust (Mosher, 1998). Proterozoic rocks in the Llano uplift comprise three distinct tectonostratigraphic terranes, each with a wide and partly overlapping range of ages encompassing the period 1.326–1.098 Ga. As in the Transition Zone of Arizona, these terranes are separated from each other by faults or shear zones. In the southern Llano uplift, the Coal Creek terrane consists mainly of plutonic rocks, interpreted as the basement of a volcanic arc. Some of the included mafic rocks are similar in composition to modern island arc and ocean-floor basalts. The Coal Creek terrane includes a large tabular body of serpentinized harzburgite (Fig. 1.18), which was tectonically emplaced within the plutonic complex.

North of the Coal Creek terrane, and separated from it by a southward-dipping ductile thrust zone, is the Packsaddle terrane. In contrast with the Coal Creek terrane, the Packsaddle terrane comprises mainly

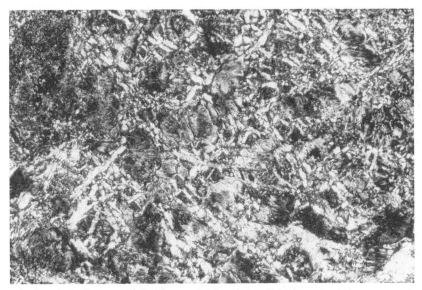

Fig. 1.18. Photomicrograph of the Coal Creek serpentinized harzburgite, from the Llano uplift of Texas. Harzburgite is a rock consisting mainly of olivine and orthopyroxene, and for the Coal Creek body is inferred to represent part of the upper mantle. Its alteration to serpentinite, which consists mainly of an aggregate of the minerals 'antigorite' and 'chrysotile,' results from pervasive alteration of the harzburgite at elevated pressure and temperature in the presence of H_2O-bearing fluids. In addition to antigorite and chrysotile, other minerals such as talc and magnesite may be present. The mesh texture, characteristic of serpentinite, is visible in this view. Crossed polarizing prisms. Field of view is approximately 1.7 mm × 2.6 mm.

metamorphosed volcanic and sedimentary strata, and includes only minor volumes of intrusive rocks. The northernmost terrane is the Valley Spring, separated from the Packsaddle by another southward-dipping ductile thrust zone. This terrane consists of metasedimentary and metavolcanic strata, ranging from mafic to felsic in composition. The Valley Spring domain may represent the southern margin of Laurentia (Mosher, 1998). Altogether, the terranes exposed in the Llano uplift are interpreted to suggest the collision of an island arc and of a trailing continent with the southern margin of Laurentia. Accretion of the Coal Creek island-arc terrane to Laurentia occurred 1.150–1.119 Ga. The high pressures and temperatures

(750 °C and 1.5 GPa) recorded by the metamorphism indicate that the crust was up to 50 km thick as a result of the collisional events (Mosher, 1998).

In the Mexican state of Chihuahua, sparse outcrops of Precambrian rocks consist of amphibolite, metagranite, and gneiss dated at approximately 1.03 Gyr. The age indicates a Grenville event, compatible with ages of Proterozoic rocks in west Texas. However, because the Proterozoic rocks of Chihuahua occur as small, possibly tectonic blocks surrounded by Paleozoic sandstone and shale, many questions remain unanswered, including their source and the direction and amount of transport to their present location (Blount, 1983; Mauger *et al.*, 1983).

Thus, the Grenville event probably represented northward-directed collision of a continental block with the southern margin of Laurentia. Based on recent geologic, geochronologic, and paleomagnetic data, it is now thought that the colliding block was most likely the Amazon craton, part of the South American continent, whose western margin collided with Laurentia 1.2–1.0 Ga (Tohver *et al.*, 2002). A suture zone, perhaps one of several, is exposed in the Van Horn area, where it comprises a series of thrust sheets, as well as in the Llano uplift of central Texas, where it is defined by scattered serpentinite bodies along a major contact zone (Mosher, 1993). The Grenville orogeny completed the growth of the craton now exposed in North America, Greenland, and Scandinavia, but it also included continents that are now long gone from the western margin of North America. It represented the final assembly of the new supercontinent named 'Rodinia.'

1.8 Summary

Over a period of 200 Myr during the Paleoproterozoic and Mesoproterozoic, the crust of the Southwest, indeed of much of Laurentia, was assembled piece by piece from geological flotsam and jetsam. Formation of most of the crust in the Southwest was complete by about 1.60 Ga. The new crust was then annealed into a strong whole, and crunched again with an unknown landmass, as Rodinia took form. This much geologists can infer. But a complete and coherent picture of the numerous island arcs, of marine basins and shorelines, and of the mountain chains resulting from each collision – events that are much clearer in the Phanerozoic – will require much additional study. Both the remarkable formation of half of Laurentia over a mere 200 Myr and the equally incredible magmatic crucible that reconstituted the new crust over huge tracts of Laurentia to stabilize the new continent are unprecedented in Phanerozoic time.

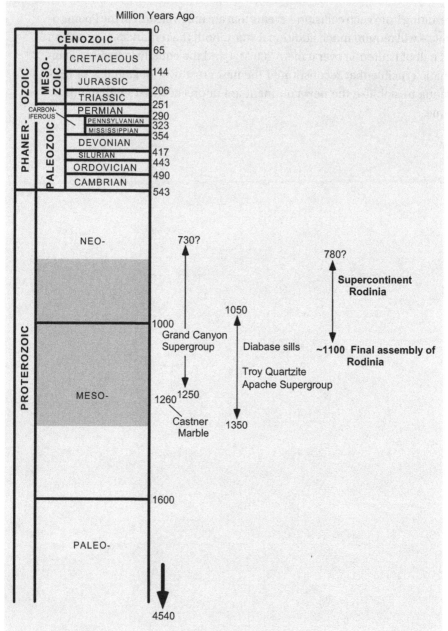

Geologic time scale. For explanation see p. 3.

2

Tantalizing time

Mesoproterozoic to Neoproterozoic

Sedimentary rocks of the upper Unkar Group, part of the Grand Canyon Supergroup (lower half of photograph), are conspicuously exposed in the Grand Canyon along the north side of the Colorado River. The Proterozoic Grand Canyon Supergroup is separated from the overlying Cambrian Tapeats Sandstone, which forms a prominent cliff just above the center of the photograph, by an angular unconformity. At this location the Unkar Group consists of the Dox Formation (variegated unit) and the interbedded Cardenas Basalt (dark unit near top of Proterozoic). View is from Lipan Point.

2.1 Mesoproterozoic and Neoproterozoic stratified rocks

Following the formation of the crust in the Southwest, completed during the Grenville orogeny 1.2–1.0 Ga, an enormous expanse of time was to pass before the widespread, relatively continuous and well-preserved rocks of the Phanerozoic Era were deposited. The present chapter covers part of that interval of time, from roughly 1.35 Ga to about 780 Ma, a length of time approximately equal to the entire Phanerozoic.

This chapter represents a considerable overlap in time with the previous chapter, but focuses on a completely different group of rocks. The rocks covered in the present chapter consist almost entirely of stratified rocks, mostly sedimentary but in part igneous, which were laid down on the previously formed crust. They rode high upon that crust and therefore escaped the great crushing stresses and high temperatures that accompanied the processes of initial crustal formation. They are somewhat arbitrarily distinguished from the little-metamorphosed stratified rocks of the previous chapter by having been deposited in intracratonic basins far from the continental margins. The beginning of this chapter at about 1.35 Ga is the earliest time for which such stratified rocks are preserved; the cutoff at 780 Ma, although it may appear to be arbitrary, signifies a major change in the tectonic events of the Southwest.

The overlap in time with the previous chapter comes about because, as discussed there, the formation of the crust in the Southwest occurred over an interval of 700 Myr, from assembly of the Yavapai province at roughly 1.7 Ga to the Grenville at about 1 Ga. Thus, undeformed sedimentary strata were deposited on Yavapai terranes at roughly 1.35 Ga in the present Grand Canyon area even as new crust was being assembled and accreted in west Texas (Grenville).

In the Southwest, major sequences of intracratonic stratified rocks of Mesoproterozoic to early Neoproterozoic age occur in three locations, the Grand Canyon of northern Arizona, the Transition Zone of central Arizona, and a broad region of southeastern California including Death Valley. They comprise parts of three, unconformity-bounded **successions** (Box 2.1) recognized throughout western North America. These successions (Fig. 2.1) are generally referred to as, from oldest to youngest, A, B, and C. However, in this book they are informally termed the pre-Grenville-, Rodinian-, and rift-to-drift successions, respectively, to emphasize their relationship

Box 2.1 **Succession** A group of rock units, usually stratified sedimentary rocks but which may include lava flows or pyroclastic units, that 'succeed' one another in chronological order. The term emphasizes that the units are ordered by time. Typically, the rock units grouped together into a single succession are close in age to each other and separated from underlying and overlying successions by a substantial time gap (unconformity).

to continental-scale tectonic events. The term 'rift-to-drift' refers to the fact that strata forming this succession record not only the rifting apart of Rodinia but also the subsequent passive margin that characterized the continent as the fragments of the former Rodinia drifted apart.

Understanding the relationships among these successions has been a significant challenge to geologists. Because of their Proterozoic age (i.e. before the Cambrian explosion of higher life forms), their paleontologic record is poor or absent, and not useful for dating. In addition, because the successions comprise mainly sedimentary rocks, they cannot be very well dated by using radiometric techniques. Moreover, each of the major successions is isolated from the others, therefore rock units cannot be traced laterally among them. Determination of **polar wander paths** (see Box 3.1) has been essential to correlation of disparate local sequences into continental-scale successions. Because the strata deposited during this time are relatively undeformed and unmetamorphosed by later events, the record is tantalizingly transparent and can be interpreted with relative ease. Local environments of deposition can be understood, transgressions and regressions of the sea are apparent, and periods of crustal extension are evident. Yet, though the time interval is long, the record of events during this interval is frustratingly incomplete, primarily because the rocks were deposited in local basins. In particular, the period of time between about 1.1 Ga and 600 Ma is not well recorded in the Southwest. Fragments of major events are recorded, yet regional correlations and the regional and global significance of these events remain uncertain.

The pre-Grenville, Rodinian, and rift-to-drift stratigraphic successions provide a broad framework for interpreting the stratigraphic and tectonic history of western North America during Mesoproterozoic and Neoproterozoic time (Elston *et al.*, 1993). Basically, the pre-Grenville succession records deposition between about 1.35 and 1.20 Ga, largely in

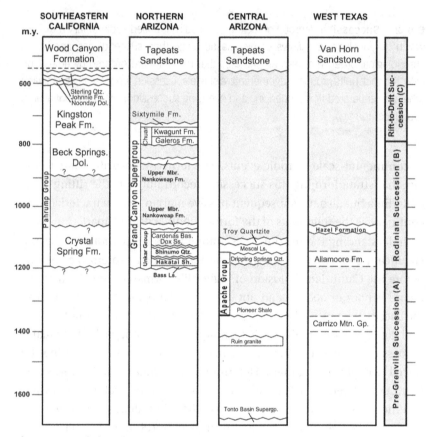

Fig. 2.1. Correlation of Mesoproterozoic and Neoproterozoic strata of the Southwest. Gray color indicates hiatuses represented by unconformities between units. Empty spaces indicate missing record or intrusive rocks not shown on diagram. The pre-Grenville-, Rodinian- and rift-to-drift successions (A, B, and C, respectively) are defined by major tectonic events (see text). The location of the Precambrian–Cambrian boundary in southeastern California sequence (dashed line) is from Corsetti and Hagadorn (2000). The age of this sequence has been adjusted to conform to inferences by Prave (1999) regarding the extension recorded by the Kingston Peak Formation. Abbreviations: Dol., dolostone; Fm., formation; Gp., group; Ls., limestone; Mbr., member; Qtz., quartzite; Sh., shale; Ss., sandstone; Supergp., supergroup. Modified from Elston *et al.* (1993) and Timmons *et al.* (2001).

isolated intracratonic basins. The unconformity marking the boundary with strata of the overlying Rodinian succession correlates to early events probably associated with the Grenville orogeny, during which the final piece of supercontinent Rodinia was assembled. Were the rocks deposited in marine or lacustrine environments? The answer is not yet fully known.

Similarly, rocks of the Rodinian succession were deposited in separate basins 1.2 Ga to 780 Ma, but lithologic data are interpreted to suggest that the basins lay on continental crust that was undergoing extension. The rocks appear to record both shallow-water marine, and terrestrial fluvial and lacustrine depositional environments. The Rodinian succession was terminated by block-faulting 800–780 Ma, recorded in British Columbia and in the Grand Canyon (Fig. 2.1).

The rift-to-drift succession postdates the 800–780 Myr-old block faulting in the Grand Canyon, but does not record quiet conditions. Rather, the succession records two episodes of continental rifting, the latter of which successfully wrested the present Southwest away from the lands to the west, forming a new continental margin. The rift-to-drift succession is discussed in the next chapter.

2.2 Intracratonic basins: the pre-Grenville succession

In central Arizona, a remarkable section of Mesoproterozoic stratified rocks, less well known than Proterozoic rocks of the Grand Canyon (next section), crops out. These strata, the Apache Group and Troy Quartzite (Figs. 2.2 and 2.3), comprise a section 850 m thick, spanning some 600 Myr and recording several cycles of deposition. The Apache Group includes rocks of both the pre-Grenville and Rodinian sequences, with a major temporal hiatus between (Fig. 2.1). The Tonto Basin Supergroup, deposited soon after crustal formation, consists of quartzose sandstone and rhyolitic volcanic rocks, whose setting is not well understood. Following a long hiatus, the lowermost unit (Pioneer Shale) of the Apache Group was deposited. Deposition began with a conglomerate, the Scanlan Conglomerate Member, a debris-flow and braided-stream deposit representing prograding alluvial fans (Wrucke, 1993). These pre-Grenville strata apparently represent local deposition in shallow, cratonic basins along or near the margin of Laurentia. Their depositional environments were partly marine,

	Diabase		
		Intrusive Contact	
	Troy Quartzite 0-365	Quartzite Member	0-150
		Chediski Ss. Mbr.	0-210
		Arkose Member	0-140
		Unconformity	
APACHE GROUP	Basalt		0-115
		Unconformity	
	Mescal Limestone 75-130	Argillite Member	0-30
		Basalt Member	0-35
		Algal Member	12-40
		Lower Member	45-82
		Unconformity	
	Dripping Spring Quartzite 140-215	Upper Member	55-130
		Middle Member	40-110
		Barnes Conglomerate Member	0-18
		Unconformity	
	Pioneer Shale 45-155	Unnamed Member	
		Scanlan Cgl. Member	0-15

Fig. 2.2. Stratigraphy of the Apache Group and associated rocks of central Arizona. Modified from Wrucke (1993).

Fig. 2.3. Mesoproterozoic metasedimentary rocks and diabase sills are exposed in the Salt River Canyon of central Arizona west of US Route 60. Symbols are as follows: uds, the upper member of the Dripping Springs Limestone; lm, am, respectively, the lower and algal members of the Mescal Limestone; tq, the Troy Quartzite. At this locality, the section is inflated about 100% by sills of diabase, which generally form slopes. See Wrucke (1989). See also Plate 3.

and perhaps they underwent periodic desiccation. Formation of basins may have been related to the postulated ongoing collisional tectonics that preceded the Grenville orogeny.

2.3 Continental extension: the Rodinian succession

The Rodinian succession records and postdates an episode of crustal extension that is widely recognized throughout the West and in the midcontinent of North America (e.g. Midcontinent rift system). The succession includes a variety of lithologies, locally coarse-grained, interpreted to have formed in separate basins. They provide evidence that the crust of Rodinia, even well into the interior, underwent a significant period or periods of extensional deformation. The lower rocks of the Rodinian succession range from shallow-water marine to tidal-flat, fluvial, playa, and even eolian. Many units are arkosic, and red beds are abundant (Hendricks and Stevenson, 1990; Wrucke, 1993). The crustal extension that resulted in these separate basins was generally synchronous with, and therefore inferred to be related to, development of the Grenville orogeny along the southeastern margin of the continent. The upper part of the Rodinian succession also includes a wide range of lithologies, but carbonates are more abundant and red beds less so. Generally, the succession records a more stable environment of deposition, which is consistent with a somewhat more stable, post-Grenville tectonic setting. These rocks postdate the Grenville orogeny by 200 Myr or more. In the Southwest, the Rodinian succession is represented by the Grand Canyon Supergroup in the Grand Canyon area, by the Apache Group and related Troy Quartzite in central Arizona, and by the Pahrump Group in southeastern California.

The Grand Canyon of the Colorado River exposes a famous and remarkable sequence of Mesoproterozoic to Neoproterozoic strata (Fig. 2.4; Box 2.2), whose importance has been recognized for more than a hundred years. Lying above the metamorphosed rocks of the Vishnu and Zoroaster Formations, they constitute the lowest stratified rocks exposed in the Canyon and one of the most complete and long-ranging Mesoproterozoic to Neoproterozoic records preserved in North America. Based on regional stratigraphic relations, the Bass Limestone appears to record marine transgression onto the crystalline basement from the west (Hendricks and Stevenson, 1990). The lowermost unit of the Bass Limestone, the Hotauta

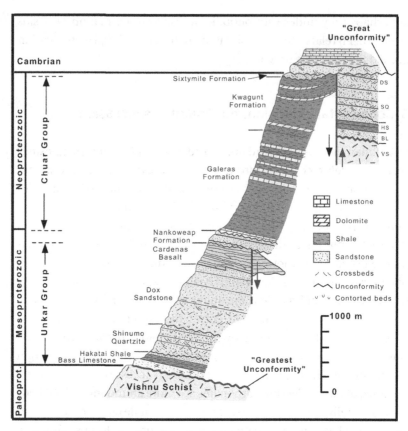

Fig. 2.4. Stratigraphic section of the Grand Canyon Supergroup of Mesoproterozoic to Neoproterozoic age, which comprises all of the sedimentary rocks in the Grand Canyon between the Vishnu Schist (crystalline basement) and the Cambrian sedimentary rocks. The Grand Canyon Supergroup records a long history of events between formation of continental crust in the Southwest and the more familiar Phanerozoic time. Stratigraphic offsets on faults are portrayed as they existed in, respectively, late Mesoproterozoic and early Paleozoic time. Abbreviations: VS, Vishnu Schist; BL, Bass Limestone; HS, Hakatai Shale; SQ, Shinumo Quartzite; DS, Dox Formation. For discussion of 'great' and 'greatest' unconformities, see Box 2.2. Modified from Elston (1989).

Conglomerate Member, is a basal conglomerate that was deposited in low areas of the Vishnu terrain as it was inundated. Clasts in the conglomerate were derived from the underlying crystalline rocks. The Bass Limestone is overlain by the (mainly?) subaqueous Hakatai Shale. The presence of intraformational breccias and conglomerates, symmetrical ripple marks, desiccation cracks, graded bedding, and stromatolites in the Bass Limestone is interpreted to indicate a low-energy intertidal to supratidal marine environment. Similar features in the lower Hakatai shale indicate a mudflat environment of deposition. Interpretations of the upper member of the Hakatai Shale range from a high-energy, shallow-marine (Hendricks and Stevenson, 1990) to a subaerial (Elston, 1993) setting.

Box 2.2 The 'great' unconformity Geology, an historical science, is preoccupied with time. Passage of time is recorded by the presence of rock units: the 'rock record.' But the absence of a record is equally important to recognize. 'Missing time' results from erosion of units or from periods of time during which no sediments were deposited. Gaps in the temporal record are referred to as unconformities, which may be manifested as layers of rock lying with an angular discordance upon older layered units: an 'angular unconformity.'

Although intervals of 'missing time' occur throughout the rock record, one of the most profound, commonly referred to as the 'great unconformity,' typically separates rocks of Precambrian age from younger rocks. Why is the top of the Precambrian in many regions marked by an unconformity? The answer lies mainly in the fact that, once continental crust is formed, it is buoyant (average density of 2.8 g/cm^3) (Rudnick and Fountain, 1995) compared with oceanic crust (average density 2.9-3.0 g/cm^3) (Fountain and Christensen, 1989) and tends to 'float' high. Continents, their cratonic interiors at least, tend to be covered by seas only during mantle events in which the ocean basins are relatively shallow and the seas displaced onto continents. This concept is discussed in later chapters. Thus, sedimentary 'cover' rocks may be deposited only infrequently. Once deposited, they may be subject to long periods of subaerial exposure and erosion, during which they are removed. Moreover, the regional tectonic events that may allow significant thicknesses of sedimentary strata to form – events such as rifting and orogenic loading in which crust is thinned, flexed downward, and displaced below sea level – tend (with prominent exceptions) to occur along continental margins. Thus, deposition and preservation of rocks in continental interiors after initial crustal formation are somewhat fortuitous.

Unconformities occur throughout the rock record of the Southwest, but most are comparatively minor. However, unconformities within and at the top of the Proterozoic are major, probably indicating crustal thickening and certainly marking the stabilization that occurred during the 1.4-Gyr-old 'anorogenic' magmatic event (Chapter 1). The younger Proterozoic sedimentary sequences such as the Grand Canyon Supergroup and the Apache and Pahrump Groups are separated from older, generally more deformed, Proterozoic rocks by major angular unconformities. The unconformity at the base of the Grand Canyon Supergroup represents about 475 Myr of missing stratigraphic record, and has been called (somewhat tongue-in-cheek) the 'greatest' unconformity. By contrast, the 'great' unconformity separating the Sixtymile Forma-tion from the overlying Cambrian sandstone represents a mere 200-Myr hiatus (Elston, 1989). Where the Grand Canyon Supergroup has been completely eroded away and the two unconformities merge, the single, 'great' unconformity represents a gap in the stratigraphic record corresponding to approximately 1.2 Gyr. In other parts of the Southwest, such as in the Sandia Mountains near Albuquerque, over 1.1 Gyr of recorded time are missing beneath Paleozoic strata.

The upper part of the Unkar Group (Shinumo Quartzite and Dox Sandstone) (Fig. 2.5) consists dominantly of fine- to coarse-grained red beds thought to have accumulated in shallow-marine, tidal-flat, and near-shore continental environments. Although not all geologists agree, it is generally thought that the formations of the Unkar Group record four cy-cles of deposition, each of which began with a marine incursion. Marine sediments were deposited first during each cycle, followed by sediments characteristic of a tidal-flat, then subaerial, environment. The lower two cycles began, respectively, with deposition of the Bass Limestone and the upper member of the Hakatai Shale; the upper two cycles began respec-tively with the lower member of the Dox Formation and the upper middle member of the Dox. The thick (985 m) Dox Formation is remarkable in that much of its thickness is characterized by features such as channels, tabular cross beds, asymmetrical ripple marks, stromatolites, mudcracks and curls, and salt casts, features suggestive of shallow-marine to subaerial tidal-flat environments. The depositional basin was filled by the end of Dox time, and the region was at or very near sea level (Hendricks and Stevenson, 1990; Elston, 1993).

Overall, the great thickness of strata and the fact that most record near-shore environments are an indication that deposition occurred in a basin that subsided at about the same rate as sediments were deposited (Hendricks and Stevenson, 1990). Yet, subsidence was not uniform during this long time interval. Rather, faulting occurred at irregular intervals, resulting in unconformities throughout the Unkar Group (Elston, 1993). What might these unconformities and other evidence for northeast-directed extension be related to? A variety of evidence suggests that the deformation recorded during and near the end of deposition of the Unkar Group was a response to crustal shortening from the southeast (Elston, 1993; Timmons *et al.*, 2001). Considering that the upper Unkar Group was

Fig. 2.5. Red quartzose, silty sandstone of the Proterozoic Dox Formation, comprising the upper part of the Unkar Group (Grand Canyon Supergroup), is exposed along the east side of Tanner Creek in the Grand Canyon. It is separated from the overlying Cambrian Tapeats Sandstone (prominent cliff-forming unit at top of photograph) by an angular unconformity. See also Plate 4.

laid down during the interval 1.20 Gyr to about 1.07 Gyr or so, it is difficult not to relate these events in some way to the major Grenville orogeny, which occurred contemporaneously along the southeast margin of Laurentia (Timmons *et al.*, 2001). Thus, these events recorded in the Grand Canyon may have been a response to the last major crust-forming event in the Southwest.

In central Arizona, strata of the Rodinian succession make up the upper units of the Apache Group (Dripping Spring Quartzite, Mescal Limestone) and the Troy Quartzite (Fig. 2.1). These are equivalent only to the Unkar Group in the Grand Canyon; strata younger than about 1.10 Gyr are not present in this region. The Dripping Spring Quartzite records shallow-water conditions, possibly with tidal flats, estuaries, and local anoxic basins. Near-sea-level conditions appear to have prevailed for a considerable period of time. The next-higher formation, the Mescal Limestone, probably accumulated in or adjacent to a shallow, intracratonic sea. Initial deposits occurred under highly saline conditions, perhaps in a supratidal environment. Strata include evaporite deposits, indications of which are the presence of hopper-shaped molds formed after halite (NaCl) crystals and pseudomorphs of dolomite after sodium-bicarbonate minerals (possibly trona or nahcolite). Algal deposits, some with a crinkly texture indicative of desiccation, and stromatolites are present near the middle of the formation. Later deposition of the algal- and stromatolite-bearing strata occurred in shallow-marine waters, but they were periodically exposed to the atmosphere, undergoing desiccation. Collapse breccias, coalesced sink holes, and silicified chert lag deposits are evidence that the Mescal Limestone was uplifted and weathered under subaerial conditions, forming a widespread karst, before subsequent units were deposited (Kenny and Knauth, 2001). Following eruptions of basaltic lava flows, the Troy Quartzite was deposited on the exposed surface. The Troy consists of a lower arkosic member deposited by fluvial and eolian processes, and an upper quartzite member of inferred marine origin (Wrucke, 1993). Together, these units are indicative of a depositional environment ranging from littoral to shallow-marine.

An important section of Mesoproterozoic to Neoproterozoic strata, spanning the age range from about 1.200 Gyr to less than 600 Myr, is also exposed in southeastern California (Fig. 2.6). None of the units from this sections correlates with the pre-Grenville succession, but the lower part (Crystal Spring Formation and Beck Spring Dolomite, part of the Pahrump Group) are equivalent to the younger Rodinian succession. The Crystal Spring Formation is approximately the equivalent of the Shinumo Quartzite and Dox Sandstone, i.e. the lower part of the succession in the Grand Canyon. From a variety of evidence, the arkosic lower part of the Crystal Spring Formation is interpreted to have been deposited during a period of extensional tectonism. Clastic debris was derived by erosion of interbasinal topographic high areas exposed above sea level. The middle and upper Crystal Spring (Fig. 2.7) and the overlying Beck Spring Dolomite (equivalent to the *upper* part of the Rodinian succession) (Fig. 2.8), are very different, consisting mainly of carbonate (principally

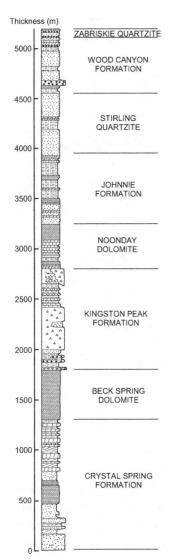

Thickness (m)

ZABRISKIE QUARTZITE

WOOD CANYON
FORMATION

STIRLING
QUARTZITE

JOHNNIE
FORMATION

NOONDAY
DOLOMITE

KINGSTON PEAK
FORMATION

BECK SPRING
DOLOMITE

CRYSTAL SPRING
FORMATION

Fig. 2.6. Stratigraphic column of the Pahrump
Group and overlying Neoproterozoic to Lower
Cambrian units in the Death Valley area. Simplified
from Link *et al.* (1993).

dolostone) and mixed clastic–carbonate strata. They record deposition in
nearshore-marine, inter- and supratidal, and fluvial depositional environments. The carbonate platform on which these units were deposited probably extended well beyond the present area of Pahrump exposure and into
areas where correlative strata remain buried or have been eroded away
(Wright and Prave, 1993).

Crystal Spring Formation   Beck Spring Dolomite

Fig. 2.7. The Crystal Spring Formation, seen in this photograph near Saratoga Spring, Death Valley National Park (California), is the lowermost formation of the Mesoproterozoic to Neoproterozoic Pahrump Group. At this exposure, the arkosic lower part of the Formation is not exposed. Massive diabase sills containing septa of stromatolitic carbonate (underlying the lower slopes across the drainage) make up the middle part of the Formation. The upper part consists of alternating beds of carbonate, quartzite, and shale/siltstone. See also Plate 5.

Supracrustal strata comprising the Rodinian succession are exposed in the Franklin Mountains and Van Horn regions of west Texas, and in the Llano uplift of central Texas. All were deposited along the southern margin of Laurentia and variously represent sediments deposited on the continental shelf or slope, in intra-arc or backarc basins, or in rifts perpendicular to the continental margin. Strata in the several regions underwent different degrees of deformation, depending on their proximity to the Grenville collisional front. In the Franklin Mountains, shallow subtidal to intertidal marine limestone (Castner Marble) (Figs. 2.9, 2.10, and 2.11) is overlain by basaltic breccia (Mundy Breccia) and by a sequence of quartzite, siltstone, and shale (Lanoria Formation). The lower part of the c. 1.26 Gyr-old Castner Marble, which was deposited on a low-energy carbonate ramp, contains abundant stromatolites (Fig. 2.10) and algal laminae (Pittenger *et al.*, 1994). The depositional environment of these shallow-marine strata

Fig. 2.8. (A) Prominent beds of orange-weathering, clay-rich dolostone and intercalated thin quartzite (darker units) of the Beck Spring Dolomite, part of the Mesoproterozoic to Neoproterozoic Pahrump Group. (B) The Beck Spring Dolomite is characterized by stromatolitic dolostone (wavy laminae in upper bed, behind scale) and interbedded quartzite (lower bed). Scale is open 21 cm. Both photographs were taken at the southern end of the Saratoga Hills near Saratoga Spring, Death Valley National Park, California. See also Plate 6A,B.

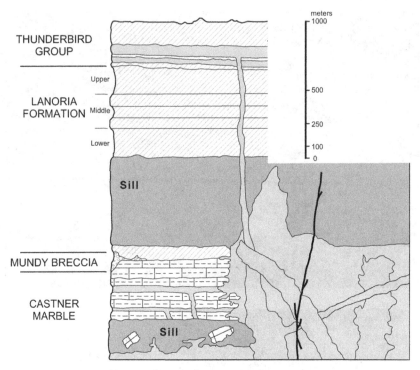

Fig. 2.9. General Proterozoic stratigraphy of the Franklin Mountains, near El Paso, Texas. Sedimentary rocks represent a sequence of shallow marine shelf sediments deposited on the southern margin of the Laurentian continent following the Mazatzal episode of crustal formation. Shaded patterns are intrusive rocks 1.13–1.15 Gyr old. Simplified from Pittenger *et al.* (1994).

is interpreted either as a continental shelf (Pittenger *et al.*, 1994) or as a shallow rift basin, formed at a high angle to the convergent margin (Bickford *et al.*, 2000). The sedimentary strata are, in turn, overlain by a sequence of silicic lava flows, ignimbrites, and tuffaceous strata (Thunderbird Group). All units were intruded approximately 1.12–1.11 Ga by granitic magmas, whose compositions *may* be indicative of an extensional tectonic environment (Bickford *et al.*, 2000).

In the Van Horn region, marine carbonate, similar to the Castner Marble, is interbedded with felsic tuff (Allamoore Formation) and with volcaniclastic sandstone, agglomerate, and basaltic flows (Tumbledown Formation). These strata, interpreted to have been deposited in a shallow-marine environment, are correlative with the Castner Marble and Mundy Breccia

Fig. 2.10. Top view of laterally linked stromatolites from the lower part of the Castner Marble, Franklin Mountains, Texas. The hemispherical layers are the limey deposits laid down by colonies of algae, which are not themselves preserved. These stromatolites formed in relatively quiet, shallow seas on a carbonate ramp along the southern margin of the Laurentian craton *c.* 1.26 Ga (Bickford *et al.*, 2000). Scale is 21 cm long.

of the Franklin Mountains. The associated volcanic and volcaniclastic strata are probably representative of either a volcanic arc adjacent to a continental margin (Mosher, 1998) or of a rift (Bickford *et al.*, 2000). In contrast with the undeformed sequence of the Franklin Mountains, the sedimentary and volcanic strata of the Van Horn area are exposed beneath a series of thrust sheets and are deformed into tight folds. Overlying the above sequence, and separated from it by the main (Streeruwitz) thrust, is a sequence of highly metamorphosed sedimentary strata, rhyolite flows, and welded ash-flow tuffs. As discussed in Chapter 1, the overlying metasedimentary and metavolcanic strata may be part of the autochthonous block of a southern, impacting continental block (Mosher, 1998) or, alternatively, of the southern margin of Laurentia (Bickford *et al.*, 2000). A synorogenic clastic wedge (Hazel Formation), consisting of more than 2.5 km

Fig. 2.11. Flat-pebble conglomerate in the upper part of the Castner Marble. The conglomerate was formed by disruption of incompletely lithified limestone laminae, possibly by storm waves, and subsequent redeposition of the clasts. Since storm-wave base may be 40–80 m deep, the conglomerate layers are interpreted to indicate slightly deeper water than that in which the stromatolites of the lower Castner Marble (Fig. 2.10) were formed (Pittenger *et al.*, 1994). Scale is 21 cm long.

of exclusively terrestrial sediments, was deposited to the north in a sedimentary basin adjacent to the developing orogen. Based on paleomagnetic and geochronologic data, the Hazel Formation is interpreted to have accumulated between about 1.10 Ga and 1.08 Ga, inferred to be the age of the deformation that created the highland source and the basin in which the Hazel Formation accumulated. The Streeruwitz thrust postdates deposition of the Hazel Formation.

An important sequence of Proterozoic metavolcanic and metasedimentary strata (including schist, marble, and quartzite) characterizes the Packsaddle terrane in the Llano uplift. These supracrustal strata are

Fig. 2.12. The light-colored, quartzo-feldspathic gneiss seen in this photograph, part of the Packsaddle Schist exposed in the Llano uplift, could have originated as a rhyolitic tuff or an arkosic sediment near an island arc (Mosher, 1998). The age of this rock is probably 1.25–1.24 Gyr. Location is 19.7 km southeast of Llano along Highway 71 (Kyle, 2000). Scale is 41 cm.

interpreted as shallow shelf and slope sediments, intercalated with mafic to felsic volcanic and volcaniclastic rocks (Fig. 2.12). The sequence may represent eroded arc materials interlayered with marine sediments along a continental margin. Similarly, in the Spring Valley terrane, which is separated from the Packsaddle terrane by a ductile thrust zone (Chapter 1), another sequence of mainly granitic gneiss, with lesser amounts of amphibolite, metagabbro, and marble, occurs. This sequence represents metamorphosed mafic to felsic volcanic and volcaniclastic strata, limestone, and arkose. The Valley Spring terrane is interpreted to represent the southern margin of Laurentia (Mosher, 1998).

2.4 Basaltic magmatism

In the Grand Canyon, a sequence of basaltic lava flows 300 m thick, known as the Cardenas Basalt, overlies sedimentary strata of the Unkar Group.

Using Rb–Sr and K–Ar radiometric techniques, the Cardenas Basalt has been fairly reliably dated at 1103–1070 Myr (Larson *et al.*, 1994; Timmons *et al.*, 2001). The sequence comprises some 10–12 flows, many of which are separated by interbedded sandstone layers. In contrast with the underlying sedimentary strata, which were deposited at least partly in a subaqeous environment, the lava flows appear to have been erupted entirely subaerially. Flows of the Cardenas Basalt are probably related to sills and dikes, which intruded all formations of the Unkar Group below the Cardenas Basalt. The sills range up to 300 m in thickness.

Approximately contemporaneously with eruption of the Cardenas lava flows in the Grand Canyon, mafic rocks were also erupted in central Arizona. Here, a sequence of basalt flows was emplaced following deposition of carbonate strata and prior to deposition of an argillite unit that, together, comprise the Mescal Limestone (Fig. 2.3). Later, mafic magma was intruded into strata of the Apache Group and Troy Quartzite, forming composite sills up to 400 m thick (Fig. 2.3). Over large areas of outcrop, the aggregate volume of the intrusive mafic rocks is as great as the combined volume of the host Apache and Troy strata (Wrucke, 1993). Ages of the sills range from 1.12 to about 1.04 Gyr, making them generally correlative with the Cardenas flows (Wrucke, 1993). Thus, it is likely that the magmatism in both the Grand Canyon and central Arizona areas is related to the Grenville orogeny.

The shallow depth of these thick and laterally extensive basaltic sills within stratified rocks close to the Earth's surface raises some interesting questions. Basaltic magmas, which are generated in the mantle, are much less dense than the mantle (approximately 2.71–2.73 g/cm^3 compared with average upper mantle density of 3.2–3.3 g/cm^3). Therefore, they must, in general, rise buoyantly toward the surface. They are quite capable (obviously) of traversing the entire thickness of the continental crust and erupting on the surface. Some of the more dramatic volcanic features of the Southwest, such as the lava fields of El Malpais in New Mexico (Chapter 8), owe their origins to basaltic melts. Basaltic volcanism is thought to be facilitated by lithospheric extension, which thins the crust and creates fractures and/or weaknesses in it. However, if basaltic melts are not able to penetrate the entire crust, then they will reach neutral buoyancy either at the base of the continental crust or possibly within the middle crust (Glazner and Ussler, 1988). The lower crust may be difficult to penetrate because its density is similar to that of the basaltic melts (2.7–2.9 g/cm^3, depending on its composition) and it is ductile compared with the mantle. In the middle crust, another density interface inhibits the ascent of basaltic magmas, a result of the compositional difference between mafic to intermediate lower crust and silicic upper crust ($\leq$ 2.7 g/cm^3). Yet the voluminous basaltic melts represented by sills in the Apache–Troy sequence ascended through most of the crust, only to stall out a short distance beneath the surface.

Although the volume of basaltic rocks of the Cardenas Basalt and related intrusive rocks, and of the diabase sills of the Apache Group – Troy Quartzite sequence, is not particularly great (*c.* 60 km^3 for the Cardenas Basalt; not estimated for the Apache–Troy sequence), they are important in representing a widespread magmatic event dated at around 1.1 Gyr (Fig. 2.13). This generally basaltic magmatic event was essentially contemporaneous with the welding of the Grenville orogen to Laurentia, discussed in the previous chapter.

The fact that much of the basaltic magmatism associated with this 1.1-Gyr event was erupted in, or at least preserved in, sedimentary basins has been interpreted to suggest that the magmatism accompanied a widespread episode of aborted intracratonic rifting. But how would a collisional orogen, such as the Grenville, give rise to such widespread rifting? One possibility is that the magmatism is the result of a plume: buoyant, upwelling mantle impinging on the base of the lithosphere. However, plumes have no particular association with convergent margins, therefore the presence of a plume closely associated with the Grenville event would be somewhat fortuitous. In addition, the conventional notion of plumes is that their radius is only about 1000 km, yet the region affected by the 1.1-Gyr old magmatic event spans a greater region. Finally, plumes are only one of several potential causes of rifting of the lithosphere, so the presence of intracratonic basins does not in any sense require a plume origin. A second possibility is that the amalgamation of a supercontinent such as formed after the Grenville collision might insulate the underlying mantle, trapping heat that would otherwise escape from the Earth's interior. The result could be upwarp of geotherms in the continental lithospheric mantle, upwelling of the mantle, and large-scale decompression melting. Again, this explanation would not imply continental rifting. Finally, a third possibility is suggested by the modern collision of the Indian subcontinent with the Asian continent. Not only has collision of these two continental masses

Fig. 2.13. Magmatism in Arizona approximately 1.10 Ga, represented mainly by diabase sills and dikes, is part of the Upper Apache Group/Troy Quartzite exposed throughout central and southern Arizona, and of the Unkar Group of the Grand Canyon in northern Arizona. In inset are shown areas of equivalent stratified rocks of the pre-Grenville succession, some of which include similar magmatic rocks. Solid lines enclose outcrop areas; dashed lines are inferred boundaries of depositional basins. Modified from Wrucke (1989) and Link *et al.* (1993).

raised the Himalaya Mountains, uplifted Tibet, and thickened the underlying crust, but, more germane to this discussion, it has resulted in tensional stresses perpendicular to the collisional direction. Thus, rift basins (including the Baikal rift) formed over a wide region of China and Siberia. These basins may be analogous to those in which the Mesoproterozoic rocks were deposited. Needless to say, issues raised by this 1.1-Gyr old magmatic event will continue to undergo scrutiny.

2.5 Timing and setting

What do these separate outcrop areas represent, and how do they relate? What orogenies do they represent? As discussed, Mesoproterozoic to Neoproterozoic strata were deposited in separate, diachronous intracratonic (perhaps fault-bounded) basins on older crystalline basement. Individual formations or groups of formations do not correlate from basin to basin, therefore they do not comprise isolated remnants of once continuous strata and are not parts of a single passive margin. Three major successions of rock units bounded by major unconformities can be discerned, however. The first (pre-Grenville succession) was deposited prior to the final assembly of the supercontinent Rodinia. The rocks record numerous periods of tectonic deformation. The second (Rodinian succession) records the tectonic upheavals associated with the Grenville orogeny followed by the relatively stable existence of the supercontinent Rodinia from the time it was amalgamated until its breakup about 780 myr. ago. The rocks preserve the record of geological events in the interior of Rodinia over the lifetime of the supercontinent. The third succession (rift-to-drift succession) is described in the next chapter.

Although substantial records of parts of Mesoproterozoic to Neoproterozoic time are preserved in special areas, the record overall is rather paltry. We should keep in mind that the lengths of time represented by the pre-Grenville and Rodinian successions (500 and 420 Myr, respectively) are very close to the length of time represented by all of the Paleozoic, Mesozoic and Cenozoic together (543 Myr). Therefore, at best, our interpretation of Mesoproterozoic to Neoproterozoic events can only be sketchy.

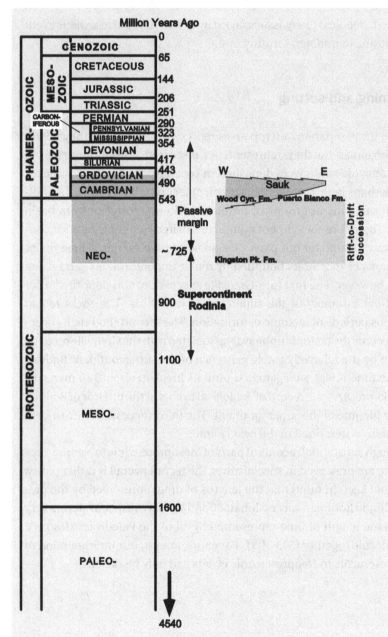

Geologic time scale. For explanation see p. 3.

The international connection: breakup of a supercontinent

Neoproterozoic to Early Ordovician

The Cambrian Muav Limestone (foreground), deposited during the maximum extent of the Sauk transgression, forms resistant cliffs above the Bright Angel Shale throughout the Grand Canyon. This outcrop is located along the Tanner Trail in the eastern part of Grand Canyon National Park.

3.1 Introduction: supercontinent Rodinia

From a variety of evidence, including the global pattern of Proterozoic rocks, geologists now recognize that Mesoproterozoic to Neoproterozoic Laurentia was actually part of a supercontinent, i.e. a 'continent' comprising continental-scale crustal blocks assembled separately and brought together by plate-tectonic processes. The Southwest and all of present North America lay interior to this vast supercontinent. The present chapter describes the demise of this supercontinent by rifting, and the transition from rifting to drifting to form a passive margin. Part of the evidence that a continent once larger than the present North America existed is the abrupt westward termination of southwest-trending Paleoproterozoic to Mesoproterozoic age provinces (Fig. 1.3). Much of the evidence is provided by Neoproterozoic strata, which are interpreted to indicate a west-facing passive margin with no sign of a continent to the west. Because Mesoproterozoic rocks of the western USA and Canada (including those of the Grand Canyon Supergroup) are interpreted as intracratonic in their setting, i.e. deposited within the interior of a continent, formation of a passive margin implies that such a continent was subsequently rifted apart. Some contention exists concerning priorities for naming the Mesoproterozoic to Neoproterozoic supercontinent (Young, 1995), but 'Rodinia' seems to be the most accepted. In addition, the timing of breakup is not as well determined as geologists might like. Very possibly a second supercontinent – Pannotia – formed as the pieces of Rodinia recollided in a different configuration. It formed for a geologically brief period of time in the latest Proterozoic to earliest Cambrian following breakup of Rodinia. What did Rodinia look like? What continent adjoined the Southwest?

Over the past 30 years, several ideas have emerged regarding the continental landmass that once adjoined the western United States. An older idea, recently restated (Sears and Price, 2000; Piper, 2000), postulated that the Siberian craton lay adjacent to the western margin of Laurentia. This continent has been termed 'Paleopangea.' Evidence includes possible correlation of source terranes on Siberia with sedimentary sequences on Laurentia, equivalent ages of mafic sills on Laurentia and Siberia, similarity of Neoproterozoic sedimentary sequences on the margins of both cratons (Sears and Price, 2000), and a good fit to the paleomagnetic data (Piper,

Fig. 3.1. One proposed reconstruction of the supercontinent Rodinia. In this configuration (called SWEAT, for Southwest US – East Antarctica), eastern Antarctica is juxtaposed against southwestern North America, and Australia lies against western Canada. The gray pattern indicates Grenville orogenic belts, formed during the assembly of Rodinia (see text). Modified from Weil *et al.* (1998).

2000). At present, the bulk of evidence does not seem to favor this reconstruction.

More recently, two other major reconstructions ('Rodinia'), both based on a variety of evidence, have emerged (Karlstrom *et al.*, 1999). Both reconstructions are similar in postulating that Australia and Antarctica, joined to each other and possibly to part of China, lay adjacent to present North America. In the first reconstruction, given the acronym SWEAT for 'Southwest US – East Antarctica,' the western USA is matched to Antarctica and western Canada is matched to Australia (Fig. 3.1). Evidence for this

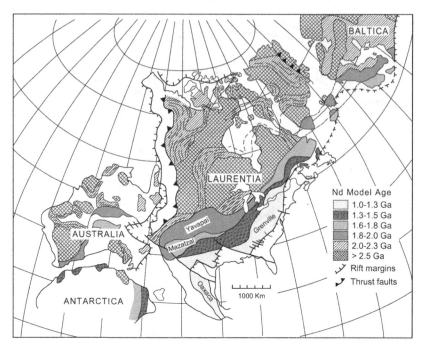

Fig. 3.2. An alternative reconstruction for supercontinent Rodinia, called AUSWUS (for Australia – Southwest US) is shown in this figure. In this geometry, Australia is juxtaposed against southwestern North America, with Antarctica lying somewhere farther south. Modified from Karlstrom *et al.* (1999).

configuration includes a postulated continuation of the Grenville orogen into the Weddell Sea area of Antarctica, correlation of Neoproterozoic sequences between Australia and Canada, and similarity of apparent polar wander paths between North America and Australia (Idnum and Giddings, 1995). An alternative reconstruction, termed AUSWUS for 'Australia – Southwest US,' places Australia adjacent to the western USA, and Antarctica farther south relative to North America (Fig. 3.2). Evidence for the AUSWUS reconstruction includes matching of inferred rift-transform segments, lineaments, and major geological provinces including juvenile Proterozoic orogens (Yavapai and Mazatzal) between Australia and the western USA (Karlstrom *et al.*, 1999; Burrett and Berry, 2000). Both models have strengths and weaknesses, and other variations between these two reconstructions exist as well (Li *et al.*, 2002). Much additional, and *interdisciplinary*, research will be required to more accurately reconstruct the former supercontinent of Rodinia in either of its forms, or its original alternative Paleopangea. This work will be international in scope, and will particularly have to develop data sets in areas of the world much less studied than the southwestern United States.

Where on Earth was Rodinia located, and how can we tell? A modern tool for tracking motion of continents is provided by **paleomagnetism** (Box 3.1). Paleomagnetic studies provide information not only on the timing of breakup, but also on the location of Rodinia relative to the Earth's poles. We know, for example, that the latest Proterozoic equator extended more or less through what is now the southwestern USA (Fig. 3.3). It cannot provide longitudinal information, of course, because coordinates of longitude are arbitrarily fixed with respect to a single geographical point (Greenwich, England) on a drifting plate (Eurasian plate). Over its lifetime of more than 300 Myr, Rodinia probably drifted through some 160° of latitude (Li *et al.*, 1995).

Box 3.1 Paleomagnetism and apparent polar wandering Magnetic minerals in certain igneous and sedimentary rocks may 'lock in' the position of the Earth's magnetic field. This magnetic signature may remain unaltered, even if the position of the rocks relative to the Earth's magnetic field should change or if the field should drift or even reverse its polarity. Paleomagnetism, the study of the magnetic signatures recorded in the rocks, is a powerful technique for reconstructing the paleolatitude of continents and for dating otherwise undatable rocks.

If, for example, strata are deposited on a continent, initially located near either pole, over a time interval during which the continent drifts toward the equator, the inclination (deviation of the magnetic field lines from the horizontal) recorded by the rock sequence will be systematically shallower in the younger units. If the continent were assumed fixed with respect to latitude, then the change in inclination would appear as a drifting, or wandering, of the pole ("apparent" polar-wander path), and observations of identical time intervals from different continents would be expected to yield identical apparent-pole positions. Yet, although more than a century's observations of the Earth's magnetic field confirm that the magnetic poles do wander slightly with respect to the Earth's rotational axis, major shifts in the Earth's apparent paleopole locations result from changes in plate locations, as indicated by the fact that polar-wander paths on different plates are different. Thus, for periods of time for which paleomagnetic data are available, paleolatitudes can be precisely reconstructed. Radiometric dating of measured magmatic rocks allows calibration of apparent-polar-wander paths. In contrast, because the Earth's magnetic field is roughly symmetrical with respect to the Earth's rotational axis, paleolongitudes cannot be reconstructed. Thus, an important application of apparent-polar-wander paths is the dating of continental separation. Two continents, formerly contiguous, describe different apparent-polar-wander paths as a function of their separate paleolatitudes. The times at which their apparent paleopoles coincide (Figs. 3.3 and 3.4) are the times at which they were in contact.

The Earth's magnetic field reverses at irregular intervals of from about 20,000 to 700,000 years, with the north pole flipping to become the south pole and vice versa. Paleomagnetic dating of continuous stratigraphic sequences is accomplished by comparing patterns of reversals to a well dated, composite reversal time scale, painstakingly assembled over decades from many dated stratigraphic sequences throughout the world, much as tree-ring dating is done.

Although the breakup of Rodinia occurred in the Neoproterozoic, this chapter ends after a major depositional cycle that was in progress during the breakup (discussed later in this chapter). This sedimentary cycle ended during the middle part of the Ordovician Period.

3.2 Breakup of Rodinia: evidence from paleomagnetism

When did Rodinia begin to break up and how can we tell? Because this book is focussed on the Southwest, we will discuss only rifting of Australia–Antarctica from the remainder of Rodinia to form the western margin of Laurentia. Evidence exists for at least two periods of Neoproterozoic rifting along what became, after breakup of Rodinia, the western margin of Laurentia (Park et al., 1995), but many uncertainties concerning this breakup remain. Primary information regarding the timing of the breakup is provided by paleomagnetic data, most of which is derived from areas outside of the Southwest.

Much insight into the timing of the breakup is derived by comparing the apparent-polar-wander paths (see Box 3.1) from North America (part of Laurentia) with those from Australia, Antarctica, and India. These latter three continents all formed a single part of Rodinia adjacent to the western margin of what became, after breakup, Laurentia, and themselves resisted breaking apart for another 250–300 Myr or so. The apparent-polar-wander path from the reconstructed continent of Australia–Antarctica–India overlaps with that from Laurentia at about 1050 Ma, but had diverged

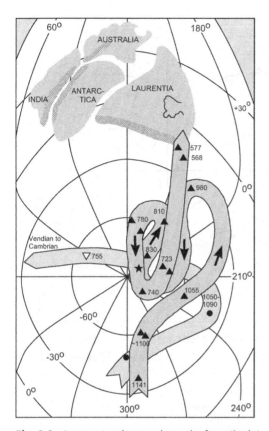

Fig. 3.3. Apparent-polar-wander paths from the latest Mesoproterozoic to the early Cambrian for Laurentia (triangles) and Australia (circles), with continents rotated into the Rodinian configuration (SWEAT configuration, see text) of Powell *et al*. (1993). Pole positions for Laurentia and Australia at different times are indicated, respectively, by triangles and circles. Ages are in Ma. An interpolated pole position for Laurentia at 755 Ma is shown by a star. Pole position for Australia at 755 Ma, inferred from the Mundine Well dyke swarm, is shown by an inverted, open triangle. The fact that the two pole positions lie approximately 30° apart is interpreted to indicate that (assuming the reconstruction of Rodinia is correct) Laurentia and Australia were separated by 755 Ma. That is, the breakup of Rodinia occurred before 755 Ma, and thereafter, both landmasses drifted along separate paths toward lower latitudes. If the SWEAT configuration of Powell *et al*. (1993) is not correct, then this paleomagnetic analysis is not relevant. Hatched areas of the continents indicate the approximate extent of Grenville-age orogenic belts. The Vendian is the latter part of the Neoproterozoic, approximately 580–543 Ma. Arrows on apparent-polar-wander path for Laurentia indicate the direction of younging. Simplified from Wingate and Giddings (2000).

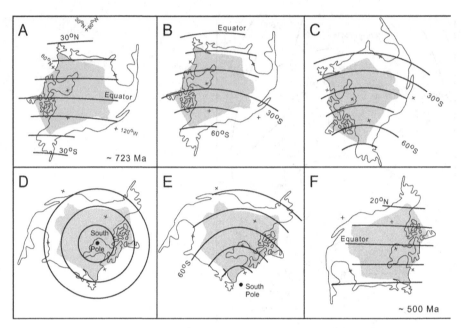

Fig. 3.4. Paleolatitude maps for Laurentia (shaded pattern) from the Neoproterozoic (approximately 726 Ma) to the late Cambrian (approximately 500 Ma), showing its rotation and possible drift over the South Pole. Lines of latitude, labeled in large font, are inferred paleolatitudes. Latitudes and longitudes in small font (panel A, not repeated in subsequent panels) are present coordinates. Modified from Park (1994).

by 755 Ma (assuming the SWEAT configuration of Rodinia) (Powell *et al.*, 1993; Wingate and Giddings, 2000). The coincidence of paleopole locations is interpreted to indicate that the two landmasses were juxtaposed in the supercontinent of Rodinia for almost 300 m.y. However, by 755 Ma the apparent-polar-wander paths diverge, indicating that the continent Australia–Antarctica–India had separated from Laurentia (Fig. 3.3). Thereafter, Australia–Antarctica–India remained at low latitudes for the rest of the Neoproterozoic. In contrast, Laurentia drifted to higher latitudes, possibly sliding right over the South Pole (Fig. 3.4). While many uncertainties remain in this broad picture, primarily owing to uncertainties in the absolute ages of some of the rocks providing paleomagnetic control, the timing generally agrees with geologic information.

3.3 Continental rifting and breakup of Rodinia: 'rift-to-drift' succession

In the previous chapter, the continent-wide Proterozoic successions (pre-Grenville, Rodinian, and rift-to-drift) were introduced and the significance of the first two described. The last of these great successions, the rift-to-drift succession, is preserved in only a few areas of the Southwest. Exposures in Nevada and southeastern California (particularly Death Valley), in the Grand Canyon, and in northern Mexico, are representative. Despite their small outcrop area, the strata of the rift-to-drift succession are very important to understanding the geologic evolution of the Southwest because they postdate much of the Proterozoic strata from other areas of the Southwest, such as in the Transition Zone of central Arizona (Troy–Apache sequence) and exposures of west and central Texas (Fig. 2.1). Thus, spanning the age range from roughly 800 to 560 Ma (Elston *et al.*, 1993; Timmons *et al.*, 2001), the strata of the rift-to-drift succession fill a critically important gap in time at the end of the Proterozoic. Rocks from Death Valley and adjacent regions of southeastern California and Nevada are particularly important in that they preserve evidence for intermittent crustal extension, which generally correlates with the breakup of Rodinia. However, in contrast with the simplistic and global view provided by the paleomagnetic data, the rocks themselves provide a less certain, and even contradictory, picture of the timing of breakup.

Proterozoic formations exposed in southeastern California and adjacent regions of Nevada span a range of ages from Mesoproterozoic to Neoproterozoic. They comprise the Pahrump Group and an overlying group of younger Proterozoic through lower Cambrian formations. Most of the strata correlate with the rift-to-drift succession and generally postdate other exposures of Proterozoic stratified rocks in the Southwest, but the precise ages and correlations of the formations remain uncertain. As described in the previous chapter, much of the lower part of the sequence, the Pahrump Group (Fig. 2.1), is part of the Rodinian succession, recording events on Rodinia. However, the uppermost part of the Pahrump Group (Kingston Peak Formation) and all of the overlying Proterozoic rocks, comprise part of the rift-to-drift succession.

Fig. 3.5. This conglomerate (diamictite) in the Kingston Peak Formation is interpreted to be glaciomarine in origin. Location is the southern end of the Saratoga Hills near Saratoga Spring, Death Valley National Park, California. Length of scale is 27 cm.

The Kingston Peak Formation is significant in recording possibly two discrete rifting events, separated by 100 Myr or more. Up to 3000 m thick, the Kingston Peak comprises coarse conglomerate, sandstone, and shale; pillow basalt and volcaniclastic units; and laminated limestone. Interestingly, it also includes deposits interpreted to be glacial in their origin (Fig. 3.5). Both the lower and upper stratigraphic units of the Kingston Peak are typified by fanglomerate, turbidite, and diamictite, rock types suggestive of rapid deposition in extensional basins, and by tholeiitic basalt. In addition, abrupt local variations in lithology and lateral facies changes are interpreted to suggest that syndepositional faults exerted strong control on basin geometry. In some cases, buried faults themselves are exhumed. These two rift successions are inferred to record crustal extension approximately 700 and 600 Ma, respectively (Prave, 1999). Locally, the younger period of extension, with consequent coarse-grained basinal deposits, persisted into the time during which the overlying Noonday Dolomite (and

Fig. 3.6. Laminae in the lower part of the Noonday Dolomite are interpreted as algal structures. Overall, the dolomitic facies of the Noonday is indicative of a shallow-water platform deposit, in contrast with its coarser-grained, basinal equivalent, the Ibex Formation. Location is in the southern Nopah Range, California (Troxel, 1974; Williams *et al.*, 1974). Scale is open 51 cm.

correlative Ibex Formation) was deposited, but overall the Noonday (an algal and clastic dolostone) (Fig. 3.6) records a transition to a shallow-marine carbonate platform (Link *et al.*, 1993; Wright and Prave, 1993). Although early work suggested that the Kingston Peak and related strata were deposited in a narrow basin oriented obliquely to the overall margin, recent recognition of Neoproterozoic erosion and large amounts of Cenozoic extension in the region render it difficult to reconstruct the geometry of the original basin.

The rift-to-drift succession also includes the Chuar Group (or most of it, at least), and the Sixtymile Formation (Fig. 2.4). The Chuar Group is composed mainly of fine-grained, thinly bedded shale and mudstone, with minor sandstone, red beds, chert, and stromatolite-bearing dolostone. In contrast with the underlying Unkar Group, red beds are not abundant. Strata of the Chuar Group primarily reflect shallow-water sedimentation with intermittent desiccation. Generally, they indicate a tectonically quiescent, shallow-marine to intertidal depositional environment (Ford, 1990). Various lines of evidence, including intraformational faults, the shape of the depositional basin, sedimentary structures, and soft-sediment deformation, record a significant extensional history during deposition of the Chuar Group (Timmons *et al.*, 2001).

Capping the Grand Canyon Supergroup, the Sixtymile Formation is a red-bed unit 60 m thick. In contrast to the underlying Chuar Group, the Sixtymile Formation is composed largely of intraformational breccia and coarse sandstone, with only subordinate siltstone and mudstone, shed from developing fault scarps that produced a block-faulted terrane reminiscent of the modern Basin and Range Province (Ford, 1990; Elston, 1993). The crustal extension recorded by the Sixtymile Formation has been called the Grand Canyon 'disturbance,' or orogeny. However, recent work (Timmons *et al.*, 2001) proposed that, rather than occurring just as a short burst during deposition of the Sixtymile Formation, deformation extended over a longer (*c.* 60 m.y.) period of time that included deposition of the Chuar Group.

Thus, together, the Chuar Group – Sixtymile Formation, and the Kingston Peak Formation and overlying Ibex Formation, preserve an important record of crustal extension and formation of basins. Was this the big extensional event, i.e. the episode of crustal extension that heralded the breakup of Rodinia? Several pieces of evidence suggest that it was. First, as indicated in the previous section, apparent paleopole positions of Laurentia began to diverge from those of Australia–Antarctica–India (East Gondwana) after about 720 Ma, suggesting that the breakup of Rodinia occurred at this time. Second, the carbonate to siliciclastic lithologies of the overlying Noonday Dolomite and younger formations (next section), in conjunction with facies relationships and paleogeography, are consistent with their deposition in a widespread, miogeoclinal environment (Wright and Prave, 1993), the newly formed western edge of Laurentia.

3.4 Passive margin: Cordilleran miogeocline

Formations overlying the Kingston Peak Formation (Fig. 2.6, 3.7) record a very different, and more quiescent, story. Generally, they consist of quartzose sandstone and mixed siliciclastic and carbonate strata that indicate

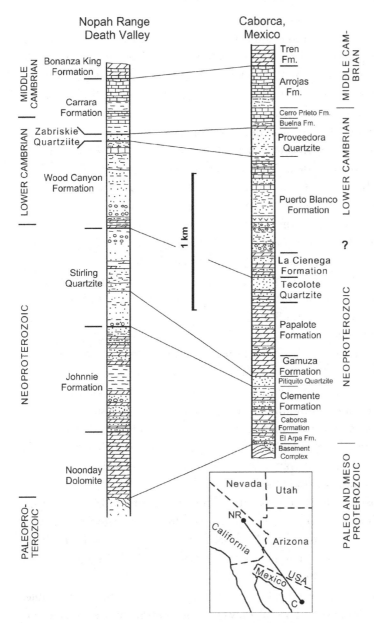

Fig. 3.7. Comparison of Mesoproterozoic to Cambrian rocks from the Nopah Range, near Death Valley, California, and the Caborca region of Mexico. Locations of sections are shown in Index map. NR, Nopah Range; C, Caborca. From Link *et al.* (1993). Position of Precambrian–Cambrian boundary within the Death Valley sequence is from Corsetti and Hagadorn (2000).

deposition in nearshore marine and fluvial environments. The Noonday Dolomite, with both basinal and platform facies, is described above. It is interpreted to record the transition from rifting to a more stable tectonic setting ('drifting'). The Johnnie Formation consists of siltstone, quartzite, dolostone, and limestone deposited in tidal and shallow marine environments. The overlying Stirling Quartzite consists mainly of fine-to medium-grained, cross-stratified sandstone, with siltstone and carbonate. Deposition of the Stirling Quartzite probably ranged from shallow marine to fluvial or braided-delta environments.

Above the Stirling Quartzite, the Wood Canyon Formation consists of sandstone, siltstone, and dolostone. In marked contrast with the generally quiescent tectonic environments of units below and above, the middle part of the Wood Canyon Formation consists of coarse-grained arkosic sandstone deposited in westward-prograding braided fluvial systems, which are interpreted to indicate active extensional tectonism and exposure of basement. From fossil and carbon isotope evidence (Corsetti and Hagadorn, 2000), the Wood Canyon Formation is inferred to span the Proterozoic–Cambrian boundary, and is generally equivalent to the Tapeats Sandstone in Arizona. Therefore, it is technically not part of the rift-to-drift succession, but nevertheless is part of the package of rift-to-drift units. Finally, the Zabriskie Quartzite consists mainly of quartz sandstone (called 'quartzite' because it is so well indurated) with minor interbedded siltstone, interpreted to be a nearshore marine to continental braid-plain deposit. It includes abundant *Skolithos* burrows (Fig. 3.8), which probably constituted dwelling and/or temporary resting structures of suspension-feeding organisms (Middleton and Elliott, 1990).

Siliciclastic material in these upper units of the rift-to-drift succession and overlying Cambrian units was derived from the craton to the east. These latest Proterozoic to Cambrian units are widespread, occurring over thousands of square kilometers throughout eastern California and southern Nevada. Regionally uniform facies trends and predominantly west-northwest-directed paleocurrents suggest deposition along a stable, slowly subsiding shelf. Overall, they comprise a north-northwest thickening wedge away from a hingeline between the miogeocline and the craton (Wright and Prave, 1993; Link *et al.*, 1993).

Equivalent strata (i.e. units of the Proterozoic rift-to-drift succession and overlying Cambrian) are also widely scattered across northwestern

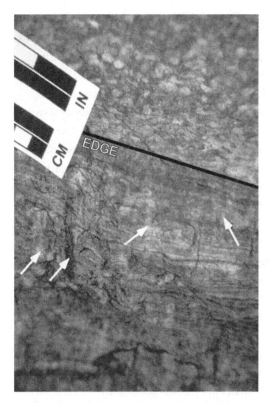

Fig. 3.8. *Skolithos* tubes are prominent in this edge-on photograph of a block of finely laminated, pinkish quartzite of the Zabriskie Quartzite (Cambrian). On the bedding-parallel surfaces (upper one-third of photo), tubes appear in cross section as white, equant shapes. Tubes are 0.25–0.5 cm in diameter. On bedding-normal surfaces (lower two-thirds of photo), tubes are exposed in longitudinal section (several representative tubes indicated by arrows). Location is Emigrant Pass across the Nopah Range near Tecopa, California (Troxel, 1974).

Mexico, with the thickest (2300 m) and most complete section (Fig. 3.7) in the Caborca area of Sonora. Generally, these represent a shelf sequence of sandstone, quartzite, siltstone, dolostone, and limestone. The lower part of the Puerto Blanco Formation, equivalent to the middle part of the Wood Canyon Formation in southeastern California, consists of volcaniclastic sandstone and boulder conglomerate, greenstone, and siltstone, suggestive of deposition in fault-controlled basins (Link *et al.*, 1993).

The significance of the extensional episode in Wood Canyon time, and particularly its relationship to the two phases of extension recorded by the Kingston Peak Formation, has been difficult to interpret. Above, we inferred that splitting of Rodinia, forming what is generally referred to as the Cordilleran miogeocline adjacent to the western margin of Laurentia, occurred near the boundary between the Rodinian and rift-to-drift successions. Evidence that deposition of the Kingston Peak Formation recorded this splitting is derived, in part, from paleomagnetism. The general westward thickening and widespread distribution of overlying platform sediments is inferred to indicate deposition on a west-facing passive margin. If this scenario is correct, then extension and basin formation recorded by the Wood Canyon (and possibly also the later extension recorded by the upper Kingston Peak? — see previous section) must represent rifting near or on the (already rifted) continental margin. This Wood Canyon rifting event has been attributed to rejuvenation of the western margin of Laurentia during rifting of its eastern margin (Powell *et al.*, 1993), but exactly why this might have occurred is uncertain. Alternatively, the Wood Canyon rifting event has been interpreted as the last of the rifting events that formed the western margin of Laurentia (see, for example, Bond *et al.*, 1984). Accordingly, the Kingston Peak events were only a preliminary, abortive rupturing event, which perhaps weakened the continent, whereas the later event reflected continued (and successful) rifting and formation of the passive Cordilleran margin, with breakup occurring 600–550 Ma. In this case, strata between the Kingston Peak and the Wood Canyon Formations represent deposition in intracratonic basins. Evidence for this latter interpretation is derived mainly from thermal-subsidence curves (Bond *et al.*, 1984). If correct, then Rodinia existed even longer than shown in the time scale at the beginning of this chapter (see also Box 3.2). In either case, the two formations together clearly record a protracted and/or episodic period of crustal extension spanning some 200 Myr or so, which is longer than continental rifts are thought to remain active. Thus, the Cordilleran margin has a complex and episodic rifting history that does not fit a simple model of continental separation (Link *et al.*, 1993). The geological evidence *per se* is not uniquely interpretable, and evidence from a variety of techniques will be required to solve this problem.

Once continental rifting 'succeeded' and the continental lithosphere was completely ruptured, sea-floor spreading began along the new plate margin. The two sides of the former continent quickly receded from each other as new oceanic crust was created, resulting in a new ocean (the

---

Box 3.2 Pannotia? Assuming that Rodinia broke apart 750–700 Ma, a second supercontinent may have existed briefly near the end of the Proterozoic. This second supercontinent has been named Pannotia (Dalziel and McMenamin, 1995; Dalziel, 1997), although the name is not accepted by all investigators (Young, 1995). If indeed Pannotia existed (and much geological investigation needs to be done to confirm its existence), then it probably represented a brief reassembly in a new configuration of parts of Rodinia before they were finally divided into the new continents of Laurentia and Gondwana. If Pannotia existed, then two supercontinents are recorded in the Neoproterozoic: Rodinia, which existed for approximately 350 Myr (from c. 1100 to 750 Ma), and Pannotia, which lasted not more than a few tens of millions of years (latest Proterozoic to early or middle Cambrian). Thus, the proposed 'super continent cycle' may be more a random agglomeration of continental fragments than a self-perpetuating, deterministic cycle (Powell *et al.*, 1993).

---

paleo-Pacific, also known as Panthalassic, Ocean) adjacent to the western edge of Laurentia. This new margin, a passive margin, existed off the western edge of Laurentia by early Cambrian. The term 'passive' refers to the fact that, once formed, it was not characterized by ongoing tectonic processes such as subduction. Instead, it simply received sediments shed from the adjacent exposed landmass. The passive western margin of Laurentia persisted for at least 200 and possibly for 375 Myr before subduction and collision was initiated along a new 'active' margin.

3.5 Transgression: the Sauk sequence

Whereas uppermost Proterozoic strata are not found, and presumably were not deposited, on the craton farther east than southeastern California and Nevada, marine strata of early Paleozoic age are widespread on the craton. In fact, much of the Phanerozoic is characterized by sequences of marine strata deposited in shallow, inland seas. These sequences mark large-scale flooding events that reached deep into the interior of the craton and persisted for time scales of tens to hundreds of millions of years. Stratigraphic sequences are bounded by interregional unconformities. In North America, stratigraphic breaks in the deposition do not correspond to those of Europe and, therefore, do not correlate well with the standard geologic periods.

The first of the major Phanerozoic **marine transgressions** (Box 3.3) is the Sauk, long recognized from both the eastern and western USA (Sloss, 1965). Beginning in the latest Proterozoic, marine waters progressively flooded the continental margin (Fig. 3.9), moving craton-ward (east and northeast in the Southwest) and marking generally increasing sea level with respect to Laurentia. The Sauk sequence is particularly well exposed in the Grand Canyon. Here, strata of Cambrian to Early Ordovician age form a classic transgressive sequence. The sequence of strata upward in the stratigraphic section represents deposits of progressively deeper water, as do time-equivalent deposits westward (Fig. 3.10).

Box 3.3 Marine transgressions: their causes? At numerous times in the Phanerozoic, the seas stood much higher relative to landmass areas than at present, flooding much of the cratons. The evidence is provided by marine shale and limestone well within the interiors of the continents. Two major sea-level highstands occurred during the Phanerozoic, the first centered on the Ordovician Period but extending from the late Cambrian to the Silurian, and the second during the Late Cretaceous. From analysis of continental flooding (Algeo and Seslavinsky, 1995), it is inferred that Paleozoic global sea level was highest during the Ordovician Period, approximately corresponding to the Sauk and Tippecanoe sequences (see Chapter 4) of Laurentia. At its maximum, global sea level was a modest 100–225 m above present sea level. The greatest flooding event of the Phanerozoic occurred during the Cretaceous (approximately coincident with the Zuni sequence), when sea level stood an estimated 175–250 m above present sea level. Sea level since the Late Cretaceous, although still flooding the outer portions of the continental crust of North America and other cratons, is lower than at most times in the Phanerozoic (Algeo and Seslavinsky, 1995). Nevertheless, cratonic seas still exist, examples of which are the North and Barents Seas. What is the cause of such major changes in sea level relative to the continents?

After ruling out local effects such as sediment loading, the question remains whether sea-level changes are eustatic or epeirogenic, or both. 'Eustatic' refers to global changes in sea level due to changes in the volume of ocean basins or in the volume of seawater; eustatic changes affect all continents simultaneously. 'Epeirogenic' refers to regional- or continent-scale changes due to tectonic causes, such as vertical displacement of the Earth's surface in response to mantle flow (known as 'dynamic topography') (Gurnis, 1992; Burgess et al., 1997; Pysklywec and Mitrovica, 1998). For example, continental margins above subduction zones subside because of downwelling of the subjacent mantle coupled to the downgoing slab (Burgess *et al.*, 1997). Epeirogenic changes in sea level would not be expected to affect all continents simultaneously.

Short-term eustatic changes in sea level may be caused by glaciation or deglaciation of continents. Long-term (first-order) changes are likely caused by changes in the depths (volumes) of the ocean basins, which are controlled by the volume of mid-ocean ridges and by the average age of oceanic crust. The depth of the ocean basins is controlled, in part, by the temperature of the crust, which is a function of its age. Temperature decreases away from the mid-ocean ridges. Because temperature controls the buoyancy of oceanic crust, the depth of the ocean floor is also a function of its age. Depth varies as the square root of age, at least for seafloor < 80 Myr old (Parsons and Sclater, 1977). Two alternative models have been proposed for significantly changing the volume of the ocean. The first model (Heller and Angevine, 1985; Heller *et al.*, 1996) relates the first-order Ordovician and Cretaceous sea level high stands to breakup of the supercontinents Rodinia and Pangea, respectively. It postulates that the volume of the ocean basins was reduced as new, thermally buoyant spreading ridges were created. The minimum volume of the ocean basins occurred 55 m.y. after breakup. No increase in overall *rates* of sea floor spreading is required.

In contrast, the second model relates first-order sea level changes to *rates* of oceanic spreading. During periods of rapid sea-floor spreading, ocean basins are younger, hence shallower. The seas tend to be displaced onto the continents, creating shallow cratonic seas. Conversely, when sea-floor-spreading is slow, the age of the oceanic crust and average basin depth increase and the seas retreat from the continents. Changes in rates of sea-floor spreading are also related to rates of convective overturning in the mantle (Gurnis, 1992; Burgess *et al.*, 1997). Thus, flooding of continents results from both epeirogenic and eustatic causes, and both are related to mantle flow. For example, increased mantle convection can lead to an increased volume of spreading ridges (eustatic control). Because plate boundaries are globally coupled, changes in spreading lead directly to changes in subduction rates and apparent sea level of the overriding continent (epeirogenic control). The Cretaceous highstand has been related to a particularly vigorous period of mantle convection and sea-floor spreading (Larson, 1991; Tatsumi *et al.*, 1998).

Possibly the extensive flooding of Laurentia that is recorded by the Sauk and Tippecanoe sequences originated mainly by the thermal subsidence of newly created passive margins that formed on both sides of Laurentia following the breakup of Rodinia, and not dominantly by rise of global sea level (Algeo and Seslavinsky, 1995). The timing with respect to the breakup of Rodinia is critical to understanding the cause or causes of these sedimentary sequences.

The limited modern flooding of the continents may be due to (1) lower spreading rates of the mid-oceanic ridges and consequent large areas of old oceanic lithosphere, and/or (2) a steep continental topography due to decreased tectonic attenuation or increased stabilization of land surface by vegetation. In addition, the modern extent of glaciation reduces sea level by about 50 m compared with an unglaciated Earth (Algeo and Seslavinsky, 1995).

The sediment deposited first as the seas transgressed was sand, which represents strand-line deposits above the exposed and erosionally truncated Proterozoic surface. In the Grand Canyon area, relief on this surface was as great as 244 m. This lower unit, the Tapeats Sandstone (Fig. 3.11), consists generally of medium- to coarse-grained, crossbedded sandstone. In the upper part of the formation, finer-grained sandstone and mudstone are more abundant. The lithology and structures of the Tapeats are generally compatible with deposition on modern tidal flats and beaches, in braided river systems, and in subtidal channels (Middleton and Elliott, 1990).

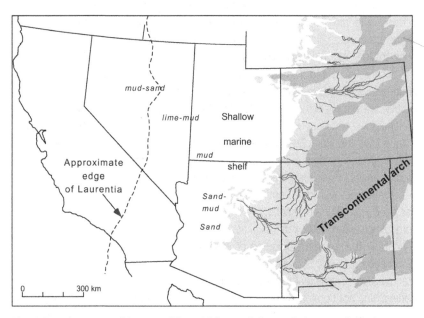

Fig. 3.9. Paleogeographic map of the middle Cambrian Period, approximately 510 Ma, during the Sauk transgression. Shaded areas are subaerial, with darker shading indicating highlands. Lines are drainages, mainly braided systems. Modified from R. C. Blakey, website: http://vishnu.glg.nau.edu/rcb/paleogeogwus.htm

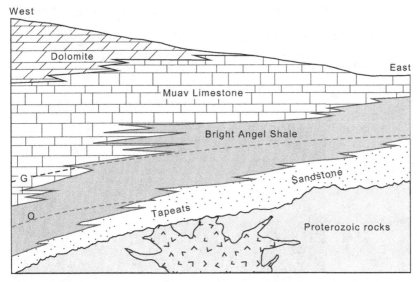

West

Dolomite

East

Muav Limestone

Bright Angel Shale

G

Sandstone

Q

Tapeats

Proterozoic rocks

Fig. 3.10. Stratigraphic correlation of Cambrian rocks exposed along the Grand Canyon of Arizona. The dashed lines labeled O and G indicate, respectively, the top of the Lower Cambrian *Olenellus* trilobite zone and the location of the middle Cambrian *Glossopleura* trilobite zone. These distinctive fossil zones are time lines. Their location in progressively shallower-water facies toward the east documents the Sauk marine transgression. This section is approximately 200 km long. From Levin (1991).

The Tapeats Sandstone can be traced both laterally and vertically into the Bright Angel Shale (Figs. 3.10, 3.11), with which it is in gradational contact. The latter formation is dominantly a greenish shale, with interbedded fine-grained sandstone and siltstone; locally, coarse-grained sandstone and conglomerate are present. Deposition was in medium to shallow water (but below fair-weather wave base) on an open shelf. Both fair-weather and storm-related processes are recognized in the Bright Angel Shale (Middleton and Elliott, 1990).

The Bright Angel Shale is in gradational contact with the overlying Muav Limestone (Chapter 3 frontispiece; Fig. 3.11). The Muav consists of thin- to thick-bedded dolomitic and calcareous mudstone and packstone, with interbedded intraformational and flat-pebble conglomerate, micaceous shale, siltstone, sandstone, and limestone. Whereas much of the Muav was deposited in a subtidal environment, laminated dolostones, interpreted

Fig. 3.11. The Cambrian Tonto Group is seen here exposed along the east side of Tanner Creek in the eastern Grand Canyon. The lower prominent cliffs are part of the Tapeats Sandstone. The slopes in the middle of the photo are underlain by the Bright Angel Shale. The thin, discontinuous cliffs above the Bright Angel are the Muav Limestone. The massive cliffs at top of photo are upheld by the Redwall Limestone of Mississippian age. The Devonian Temple Butte Formation is present between the Muav and the Redwall, but is not prominently exposed.

as cryptalgal laminations, and intraformational conglomerates may be indicative of an intertidal setting. A reasonable depositional model involves deposition of much of the Muav in deeper water far offshore, but some deposition occurred on offshore shoals or on tidal flats surrounding coastal islands (Middleton and Elliott, 1990). Together, these three Cambrian units, Tapeats Sandstone, Bright Angel Shale, and Muav Limestone, make up the Tonto Group. In the western Grand Canyon region the Muav Limestone is overlain by a sequence of dolostones (Grand Wash Dolomite). This unit was deposited in a subtidal and possibly intertidal environment of uncertain relationship to the above formations (Middleton and Elliott, 1990).

The continental platform that developed along western Laurentia wrapped around the southern margin of Laurentia into the area that is now southern Arizona, southern New Mexico, northern Mexico, and west Texas.

System	Stratigraphic unit		Thickness m	Description
Permian	Hueco Limestone		123	Dark-gray, fossiliferous limestone
	— - disconformity - —			
Mississippian	Rancheria Formation		67	Limestone, chert, and shale
	— - disconformity - —			
Devonian	Percha Shale		76	Dark-gray shale
	— - disconformity - —			
Silurian	Fusselman Dolomite		451	Alternating light and dark dolomite units
	— - disconformity - —			
Ordovician	Montoya Fm.	Cutter Member	61	Fossiliferous limestone and dolomite
		Aleman Member	43	Limestone, dolomite, and chert
		- disconformity -		
		Upham Member	15	Dark-gray, massive dolomite
		Cable Canyon Mbr.	6	Dolomitic sandstone; sandy dolomite
		— - disconformity - —		
	El Paso Fm.	Padre Member	82	Limestone with chert nodules and lenses
		McKelligon Mbr.	171	Light-gray limestone
		Jose Member	5	Thin-bedded, dark-gray, silty limestone
		Hitt Canyon Mbr.	101	Sandy dolomite, limestone in upper part
Cambrian	Bliss Formation		30-73	Arkose, arkosic sandstone; minor limestone and dolomite
	— - nonconformity - —			
	diamictite		12	Massive, red, muddy conglomerate

Fig. 3.12. Stratigraphic column of Paleozoic rocks in southwestern New Mexico. Pennsylvanian rocks are not present. From Clemons (1998).

As in the Grand Canyon area, the lower deposits of the Sauk sequence were strand-line deposits, the Bliss Sandstone. Whereas the Tapeats is lower Cambrian, the Bliss Sandstone spans a longer period of time, from the late Cambrian to Early Ordovician (Fig. 3.12). It rests unconformably or disconformably on a variety of Proterozoic lithologies (Fig. 3.13). The Bliss consists dominantly of medium- to coarse-grained glauconitic and hematitic sandstone, with minor interbedded siltstone, limestone, and dolostone. Basal beds are commonly pebbly sandstone. The Bliss Sandstone is a transgressive sandstone, deposited in shoreline and shallow-marine (subtidal to intertidal) environments. Interbedded carbonate and sandstone beds were probably laid down on tidal flats, with small pools and channels (Clemons, 1998; Mack *et al.*, 1998).

Fig. 3.13. The late Cambrian to Early Ordovician Bliss Sandstone, seen here nonconformably overlying the Proterozoic Red Bluff Granite, is similar in stratigraphic position to the Tapeats Sandstone of northern Arizona in forming the basal unit of the Sauk transgression. Whereas the Tapeats is early Cambrian in age in northern Arizona, transgression occurred later in southern New Mexico and west Texas. Location is McKelligon Canyon in the Franklin Mountains near El Paso, Texas.

However, the dominant deposits of the Sauk sequence in the southern Arizona–southern New Mexico part of the Southwest comprise a thick (up to 750 m in the southern USA) (Goldhammer *et al.*, 1993) shallow-marine carbonate section that accumulated on a gently dipping platform bounding the southern margin of Laurentia. These carbonates, the El Paso Formation, gradationally overlie the Bliss Sandstone. The El Paso consists of fossil-rich, stromatolitic limestone and dolostone (Fig. 3.14), locally containing sandy and silty beds. It was deposited dominantly in an intertidal to shallow subtidal environment. Faunal evidence and many sedimentary features are interpreted as indicating that much of the El Paso Formation was deposited in quiet, low-energy environments which were periodically disturbed by storms or strong winds. Some sedimentary structures are also interpreted as ephemeral tidal-channel deposits (Clemons, 1998;

Fig. 3.14. The Lower Ordovician El Paso Formation, deposited along the southern margin of Laurentia, preserves a rich assemblage of shelf fauna including *Nuia* (tubular organisms), trilobites, gastropods, cephalopods, echinoderms, sponges, and algae. Scale is in cm. This outcrop is near Capital Dome in the Florida Mountains, New Mexico (Clemons, 1998).

Mack *et al.*, 1998). The distribution of the El Paso Formation, and its dramatic southward thickening (Fig. 3.15), is evidence that it was deposited along an east-trending coastline. This phase of sedimentation ended by the Middle Ordovician, as the widespread Sauk seas withdrew from much of the Southwest. Previously submerged areas were left high and dry, and karstic topography developed on the platform carbonate rocks.

What was the cause of the Sauk marine transgression, observed not only in western North America but also on other continents (i.e. other margins of the former Rodinia)? While sediment loading may be part of the explanation, other possible mechanisms relate to formation of new passive margins bordering Laurentia and to a eustatic rise in sea level (see Box 3.3). A new spreading-ridge system formed as Rodinia broke apart near the end of the Proterozoic. Oceanic crust, newly formed at the ridge system, was warm, and therefore buoyant, compared with more mature oceanic crust. As a result, the overall volume of the ocean basins was reduced, displacing

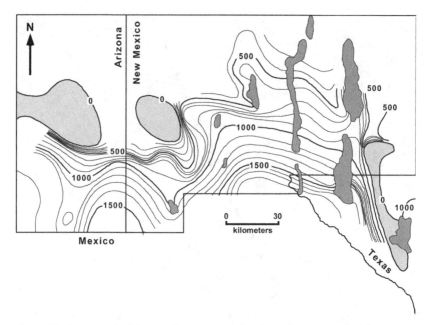

Fig. 3.15. Preserved thickness of upper Cambrian through Lower Ordovician strata (Bliss Sandstone and El Paso Formation) from southern Arizona to west Texas. Contours are in feet (= 0.3048 m). Light gray areas are early Paleozoic uplifts; dark gray areas are modern mountain ranges. From Goldhammer *et al.* (1993).

water onto the continents (Bond *et al.*, 1984). As is discussed in subsequent chapters, the Sauk was only the first of several major marine transgressions. While it did correlate with the opening of a major ocean basin, other transgressions did not.

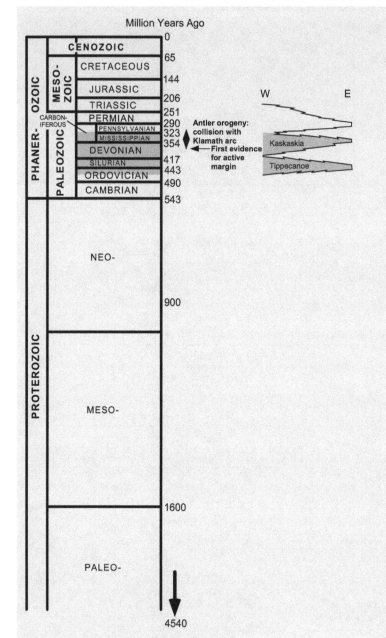

Geologic time scale. For explanation see p. 3.

Stability to orogeny: transformation of the western margin

Middle Ordovician to Pennsylvanian

Outcrops of Escabrosa Limestone of Mississippian age near Tucson, Arizona. The Escabrosa, which is equivalent to the well known Redwall Limestone of the Grand Canyon, was deposited on the continental shelf of southern Laurentia.

4.1 Introduction

This chapter commences with events of the Middle Ordovician, a somewhat arbitrary beginning point in terms of the standard geologic periods but one that makes sense with respect to marine transgressions and sequences of strata in the Southwest. It corresponds to a major unconformity at the end of the Sauk transgression, expressed over wide areas of the Colorado Plateau and indeed of much of the North American craton. In the Southwest, this unconformity and the strata immediately below and above are nowhere better observed than in the Grand Canyon, a classic section in North American geology. In other parts of the Southwest, such as western Utah, southern Arizona and New Mexico, and northern Mexico, this division is arbitrary because marine waters continued to inundate the continental margin even while more central areas became emergent.

This chapter encompasses a major tectonic change in the plate boundary along the western margin of Laurentia. Following the Late Proterozoic rifting and breakup of Rodinia, a passive margin formed along western Laurentia as the two continental fragments drifted slowly apart. Western Laurentia and Antarctica–Australia were separated by a growing ocean basin (Fig. 4.1), presumably fed by a mid-oceanic ridge system. Thus, the overall theme for the Southwest during the first approximately 100 Myr of time represented by this chapter was *stability*. Not until sometime in the Late Devonian did the first evidence for an *active* margin appear, i.e. a margin characterized by subduction and by compressional tectonics. Whether this intra-oceanic arc originated far from the margin of Laurentia or resided near the coast is unknown (Ingersoll, 1997). What initiated this change? The answer is probably long lost with the destruction of the oceanic plates of the time.

4.2 Marine transgressions and stability

The Paleoproterozoic to Mesoproterozoic crystalline crust with its fringe of Neoproterozoic and Cambrian to Lower Ordovician strata was little deformed, remaining either slightly above or slightly below sea level throughout the early and middle Paleozoic. Sedimentation during the middle Paleozoic is represented by two major cratonic **sequences** (Box 4.1), the

Fig. 4.1. The global paleogeogaphy of Laurentia in the Late Ordovician to Early Silurian. In this view, Antarctica–Australia, which were to become part of Gondwana, are hidden from view on the other side of Earth. From Mac Niocaill *et al.* (1997).

Tippecanoe and Kaskaskia (Fig. 4.2), introduced in the previous chapter. These sequences mark regional and continent-wide flooding and formation of epeirogenic seas that spread far inland, subsequently retreating and leaving much of the continent again subaerially exposed.

Box 4.1 Sequences and sequence stratigraphy Sequence stratigraphy is a methodology for correlating and interpreting the three-dimensional lithologic relationships within marine and mixed marine–non-marine strata. Building on the earlier work of Sloss (see, for example, Sloss, 1965) and others, the concept of sequence stratigraphy was developed in the 1970s by Peter Vail and other scientists at Exxon

Production Research Company as a tool for interpreting seismic reflection profiles. By then, enough seismic data existed and were of sufficient resolution that large-scale depositional geometries could be resolved, geometries that were not evident from well data alone and that were not visible at outcrop-scale except in the largest of cliff sections. Although the elements of 'sequence stratigraphy' existed long before the term was coined, the new paradigm and the initially unfamiliar terminology that accompanied it have reinvigorated the study of regional stratigraphy and paleo-environments.

The fundamental lithologic unit of sequence stratigraphy is the *sequence*, defined as a package of genetically related strata bounded above and below by regional unconformities or by their correlative basinward conformities (Fig. 4.3). The thickness of a sequence is generally measured in tens to hundreds of meters, a scale that is useful in seismic stratigraphy. A sequence is interpreted to correspond to a transgression–regression cycle of 1–5 Myr duration. Thus, the major Phanerozoic craton sequences (Sauk through Tejas), introduced in Section 3.5 (Fig. 4.2), would correspond to 'megasequences' in the terminology of sequence stratigraphy.

Boundaries between sequences comprise regional unconformities cut during a significant drop in relative sea level. As the shoreline regresses, base level of streams is lowered, incising the landward part of the basin. The resulting sedimentary record consists of valleys filled with fluvial or tidal/estuarine sediment. In the seaward parts of basins, the unconformity passes into a surface of continuous marine deposition (the correlative 'conformity'). Sequence boundaries juxtapose non-marine or shallower-marine facies (above) against deeper-marine facies (below).

On an outcrop scale, the basic building block of sequence stratigraphy is the *parasequence*, which is an upward-shallowing succession of facies on a meter to decimeter scale. Parasequences correspond to sea level changes on a much smaller scale than do the sequences. Parasequences are bounded above by *flooding surfaces*, which typically display erosional features. Above these discontinuities, rocks record a deeper-water facies. Individual parasequences are generally too thin to be resolved in seismic reflection profiles. Parasequences are stacked into systematic vertical successions, called *parasequence sets*, which may comprise successively deepening or shallowing water. The lowermost parasequence sets of a sequence mark a lowstand in sea level, and hence form what is referred to as a *lowstand systems tract*, typically deep-sea fans or shelf deposits. The lowstand systems tract comprises the sediments deposited during erosion of more landward parts of the basin, hence are correlative with sequence boundaries. Similarly, the uppermost parasequence sets within a sequence, corresponding to maximum transgression, comprise a *highstand systems tract*. Between these is an aggradational parasequence set, representing relatively constant sea level.

Although the principles and methodologies of sequence stratigraphy were largely developed from studies of siliciclastic strata, they apply just as well, with some modifications, to carbonate strata.

Sequences were originally thought to result from eustatic changes in sea level alone, hence to be correlative on all continental margins. It is now recognized that sequences result from both eustatic and epeirogenic causes (see Box 3.3).

Numerous textbooks have been written about the principles and applications of sequence stratigraphy, but for excellent short discussions see Mack *et al.* (1998), Mulholland (1998a, b, c), and Schlager (1999).

4.2.1 Tippecanoe

Shallow-marine waters of the Tippecanoe sequence inundated the southern margin of Laurentia in what is now southern Arizona and New Mexico, northern Mexico, and west Texas (Fig. 4.4). Here, the strata laid down during the Tippecanoe have remained relatively undeformed by later tectonic events, in contrast with rocks of similar age that were deposited along the western margin of Laurentia. Thus, they are illustrative of the depositional conditions associated with the transgressions. Although briefly interrupted in the Early Silurian, deposition of platform rocks was otherwise continuous from Middle or Late Ordovician to the Late Silurian (Mack *et al.*, 1998; Clemons, 1998). These platform units are represented by the Montoya Group (Middle to Late Ordovician) and Fusselman Dolomite (Middle Silurian) (Fig. 4.5). The basal formation of the Montoya Group (Cable Canyon Sandstone) records the initial transgression of the Tippecanoe marine waters over a landscape developed on subaerially exposed carbonate strata of the El Paso Group. In this regard, it is similar to the Tapeats Sandstone, which records the initial transgression of the Sauk sequence. In places over 13 m thick, the Cable Canyon Sandstone consists of cross-bedded, well-sorted sand grains in a matrix of carbonate mud. It is interpreted to represent, in part, the overriding of coastal dunes and beach deposits and their redeposition as broad, subaqueous dunes of quartz sand (a 'sandwave complex') in a subtidal, open-marine depositional environment. Extensive penecontemporaneous bioturbation by seafloor fauna, evidenced in part by sand-filled burrows, has destroyed many of the original depositional features. Modern sandwave complexes are described from open-marine and estuarine settings on broad shallow shelves in water depths of 5–60 m

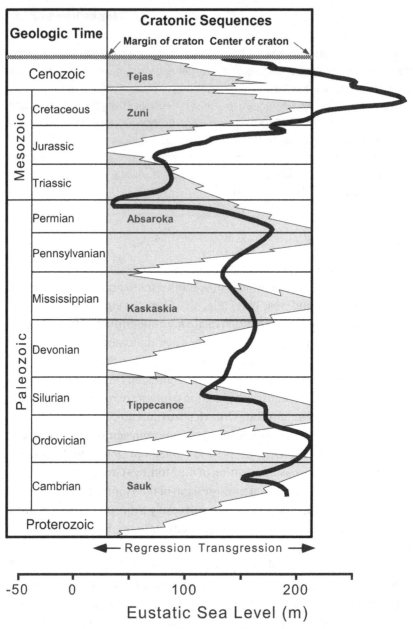

Fig. 4.2. Cratonic sequences of North America, after Sloss (1965). Areas in medium gray represent sequences of strata, separated from each other by major unconformities (white). The record of sequences is most complete near the margins of the craton. For comparison, the hypothetical eustatic sea-level curve from Algeo and Seslavinsky (1995) is shown (Box 3.3). Sea-level curve has been modified to fit the time scale of Sloss (1965), which is different from the time scale used in this book. Rock sequences are only partly controlled by eustatic sea level (see text).

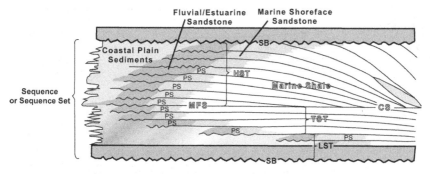

Fig. 4.3. Components of an idealized stratigraphic sequence. SB, the basal unconformity, or sequence boundary; HST, highstand systems tract; TST, transgressive systems tract; LST, lowstand systems tract; MFS, maximum flooding surface, also called the condensed section (CS); PS, parasequence. From Mulholland (1998a).

dominated by strong tidal currents. One such analogue may be the North Sea. The Cable Canyon Formation is important in that it exposes internal structures of sandwave complexes, which are difficult or impossible to study in modern complexes (Bruno and Chafetz, 1988).

The overlying formations of the Montoya Group generally consist of fossiliferous dolostone (Fig. 4.6) with abundant dark chert. They are similar to the overlying Middle Silurian Fusselman Dolomite (Fig. 4.7), which is over 450 m thick in parts of southern New Mexico (Clemons, 1998). Together, these two units record accumulation of limey muds in a shallow epicontinental sea, far from sources of terrigenous sediments.

In contrast with sedimentation along the western and southern margins of Laurentia, marine waters of the Middle Ordovician to Early Devonian Tippecanoe sequence did not reach into large parts of the continental interior. Instead, large regions of the Southwest remained above sea level through this period of time, and consequently preserve no sedimentary record. In the Grand Canyon (Fig. 4.5), rocks of Late Devonian age (Temple Butte Formation) directly overlie Cambrian rocks (Muav Formation), which were formed during the previous Sauk transgression. Hence, the entire Ordovician, Silurian, and Lower Devonian are missing from the Grand Canyon and from throughout northern Arizona (Irwin *et al.*, 1971) and New Mexico.

4.2.2 Kaskaskia

Yet again, beginning in the Early Devonian, a major marine transgression, the Kaskaskia, spread over the continent (Fig. 4.2). This time, marine waters

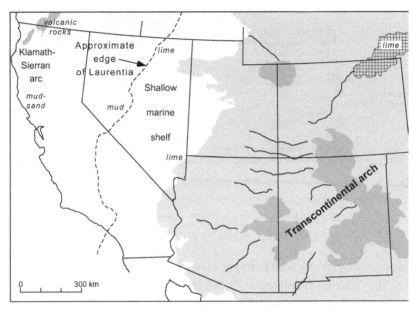

Fig. 4.4. Paleogeography of part of the western USA during the Early Silurian (approximately 430 Ma). During the maximum extent of the Tippecanoe transgression, parts of Utah and most of Nevada and California were submerged. Depositional environments ranged greatly across Nevada from shallow shelf in the east to deep basinal in the west. However, much of the Southwest remained above sea level. This period of time is marked by a major unconformity in the Grand Canyon region between the Cambrian Tapeats Sandstone and the Devonian Temple Butte Formation. The Transcontinental Arch was a broad, uplifted region of Laurentia that remained above sea level from earliest Paleozoic (and probably before) until the Pennsylvanian. Shaded areas are subaerial, with darker shading indicating highlands. Lines are drainages. Modified from R. C. Blakey, website: http://vishnu.glg.nau.edu/rcb/paleogeogwus.htm

eventually inundated most of the Southwest, remaining at least through the middle part of the Mississippian Period. This sequence left a major page in the rock record of the Southwest, which is particularly well exposed in the Grand Canyon (Fig. 4.5). The earliest unit deposited during the west-to-east transgression was the Temple Butte Formation, generally a dolostone to sandy dolostone. In many areas in and near the Grand Canyon, lower parts of the Temple Butte accumulated in channels and depressions eroded into the underlying Cambrian strata. Channels are up to 30 m deep. In places where channels are absent, Devonian strata lie

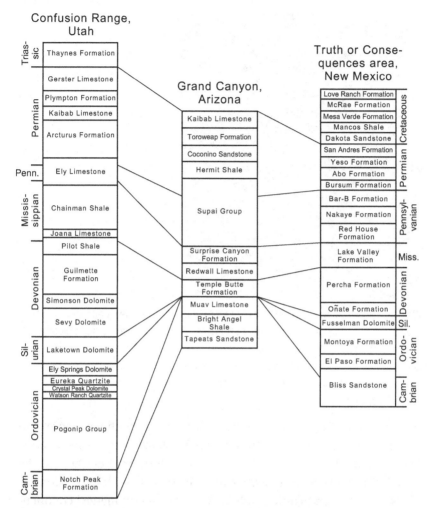

Fig. 4.5. Comparison of geologic units exposed in western Utah and in the Grand Canyon area with those of south-central New Mexico for the Paleozoic and Mesozoic Eras. In particular, rocks of the Tippecanoe sequence are present in western Utah and in southern New Mexico, but missing from northern Arizona. Modified from Hintze (1988), Beus and Morales (1990), and Clemons and Osburn (1986). These three sections represent rocks deposited during the Tippecanoe and Kaskasia sequences, and illustrate differences in depositional patterns across the Southwest.

Fig. 4.6. Photo of iron-stained brachiopod–coquina zone in the Cutter Member of the Montoya Formation (Ordovician), dominantly a cherty limestone and dolostone unit. The lithology and rich fossil record of most of the Montoya Formation are consistent with deposition on a low-energy shelf along the southern margin of Laurentia. Coin is 2.4 cm in diameter. Location is Victorio Canyon in the Florida Mountains, New Mexico (Clemons, 1998).

on Cambrian units with no angular discordance, even though more than 100 million years of time is unrepresented. Rarely, basal conglomerate beds composed of subrounded dolostone pebbles are present. Overall, the Temple Butte Formation appears to have accumulated in intertidal to shallow, subtidal, open marine conditions on a gently submerged continental shelf (Beus, 1990). Time-equivalent strata in southern New Mexico, the Oñate Formation and Percha Shale (Fig. 4.5), comprise a range of fossiliferous calcareous, argillaceous, and sandy shelf sediments, but in particular include significant thicknesses of claystone and dark shale deposited in deep basins (Clemons and Osburn, 1986).

Although generally a transgressive unit, the Temple Butte is dramatically different from other transgressive units such as the Tapeats Sandstone (Chapter 3) and the Cable Canyon Formation (above), which are dominated by quartz sandstone and which preserve evidence for high-energy littoral environments. The Temple Butte seas transgressed over a landscape of

Fig. 4.7. Algal laminae (alternating light and dark layers) in the lower dark-gray member of the Fusselman Dolomite (Silurian) are interpreted as intertidal algal mats on a stable carbonate platform (Clemons, 1998). Extensive dolomitization of the Fusselman has destroyed much of the fossil record. Location is Victorio Canyon in the Florida Mountains, New Mexico. Scale is in cm.

very low relief dominated by carbonate rocks. No significant supply of siliciclastic sediment was available.

The Kaskaskia transgression was not a single, smooth onlap and withdrawal. Instead, marine waters spread onto the continent and then withdrew several times. The Temple Butte Formation is separated from the next overlying Kaskaskia unit, the Redwall Limestone (Figs. 4.8, 4.9), by an irregular erosion surface of minor relief. Locally, the base of the Redwall is marked by a conglomerate composed of angular clasts of dolostone and limestone derived from the underlying Temple Butte Formation. The base of the Redwall is time-transgressive, being younger to the east. The Redwall Limestone, one of the most prominent units in the Grand Canyon because it forms sheer, high, red-stained cliffs, is nearly pure carbonate – limestone and dolostone – but the lower part contains abundant thin chert

Fig. 4.8. The Redwall Limestone forms massive, vertical cliffs wherever it is exposed. Here, the distinctive light and dark banded appearance, imparted by alternating chert and carbonate beds, of the Thunder Springs Member is visible along the Tanner Trail in the Grand Canyon.

beds, which probably represent silicified blue-green algal mats. The Redwall comprises two transgressive–regressive sequences, separated by another unconformity (Fig. 4.10).

The Redwall is representative of carbonate strata of the early to middle part of the Mississippian, which are widespread in the Southwest and throughout the West. In the San Juan Mountains of southwestern Colorado and in northwestern New Mexico, the Redwall is known as the Leadville Formation. Equivalent rocks in central and southern New Mexico, mainly limestone, cherty limestone, and minor shale, are known by several different

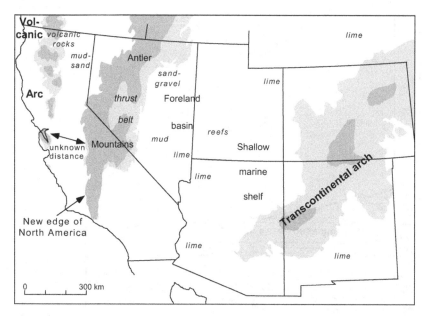

Fig. 4.9. Paleogeography of part of the western USA during the earliest Early Mississippian (approximately 358 Ma), about the time that the lower part of the Redwall Limestone (Fig. 4.8) was deposited. Shallow marine sediments were deposited throughout much of the West and Southwest, except for the region referred to as the Transcontinental Arch. Shaded areas are subaerial, with darker shading indicating highlands. Modified from R. C. Blakey, website: http://vishnu.glg.nau.edu/rcb/paleogeogwus.htm

names, including the Arroyo Peñasco Group and the Lake Valley Formation (Figs. 4.5, 4.11). In the northern Rocky Mountains the Redwall is equivalent to the well-known and widespread Madison Limestone. Overall, the Redwall and equivalent carbonate formations were deposited during the height of the Kaskaskia transgression in a shallow sea on the submerged continental shelf, under alternating transgressive and regressive conditions.

The Redwall Limestone was widely distributed on the Colorado Plateau and in the southern Rocky Mountains during the middle part of the Mississippian. However, beginning in the early Late Mississippian, the shelf seas began to withdraw. Subaerial conditions prevailed for several millions of years thereafter, during which a karstic erosion surface formed on the limestone. By the Late Mississippian, a west-draining, dendritic drainage

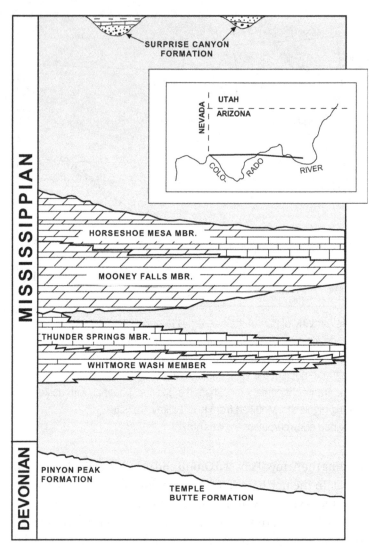

Fig. 4.10. Relative age and distribution of the Redwall Limestone across northwestern Arizona. The Redwall comprises two transgressive–regressive sequences, the first represented by the Whitmore Wash (transgressive) and Thunder Springs (regressive) Members, the second by the Mooney Falls (transgressive) and Horseshoe Mesa (regressive) Members. It is separated from the underlying and overlying units by erosional unconformities, indicating subaerial (hence regressive) conditions. These relations emphasize that the Kaskaskia sequence was not a single, simple depositional cycle. Shaded pattern indicates that no rocks are present for this time interval. Inset shows location of cross section. Modified from Beus (1990).

Fig. 4.11. Equivalent to the famous Redwall Limestone of the Grand Canyon and adjacent regions, the Mississippian Lake Valley Formation of southern New Mexico also represents limestone deposited in shallow cratonic seas during the Kaskaskia transgression. Seen in this photograph is a fold at the base of a 'Waulsortian mound' in the Lake Valley Formation. Waulsortian mounds are massive to poorly bedded mound-shaped masses of carbonate rock surrounded by steeply dipping bedded carbonate rocks. The origin of Waulsortian mounds is debated, but in the Sacramento Mountains they were most likely generated by transport of carbonate mud, in part as semicoherent masses, down the continental slope by debris flow, grain flow, and turbidity currents, with subsequent redeposition. Frictional forces caused shear on the base of the slides and flows, resulting in internal deformation and drag, as observed in this photograph (Giles, 1998). This outcrop is located in Muleshoe Canyon in the Sacramento Mountains, near Alamogordo, New Mexico.

network had developed in the Grand Canyon area, with channels up to 120 m deep. The channels were the vestiges of karst landforms, which evolved from the collapse of caves and caverns in the Redwall Limestone. Conglomerate and sandstone were deposited in the channels, karsts, and caves. Along the Cordilleran shoreline, in northern Arizona, a minor marine transgression followed. Discontinuous lenses of terrigenous–clastic and carbonate sediments were deposited in channels, interpreted as inlet lagoons. Groundwater and nearshore marine water became stratified in inlets, with fresh or brackish water overlying denser sea water (Harriet, 1992).

Collectively, these terrestrial and marine strata form the Upper Mississippian Surprise Canyon Formation. The Mississippian Period closed with subareal conditions throughout much of the Southwest. Throughout wide areas of Utah, Colorado, New Mexico, and other areas of the West, weathering of the limestone produced a rich, red soil, which filled fractures, sinkholes, and caves in the Redwall Limestone. The ancient soil, which was generally reworked in Pennsylvanian time, produced a thin unit of red, purple, and maroon mudstone and shale, which separates the Mississippian limestones from the Pennsylvanian deposits. In Utah and throughout much of the West, this reworked paleosol is known as the Molas Formation. The equivalent unit in central New Mexico is the Log Springs Formation.

Box 4.2 Meteorite impacts in the Southwest? Rocks in the Southwest contain evidence for several meteorite impacts since the Proterozoic, but typically the evidence is rather cryptic. In southern Nevada, the Late Devonian (370–355 Myr old) Alamo Breccia, near Ash Springs and Alamo, Nevada, is interpreted as recording an impact on the shallow carbonate shelf along the western margin of Laurentia. The Alamo is a single layer of limestone breccia composed of clasts ranging from sand-sized particles to blocks 80 m × 500 m in size. The lower part is a chaotically bedded debris deposit that grades upward into a turbidite. The Alamo Breccia is interpreted as a bed of debris formed within the time span of a few hours or days in the wake of a giant slide on the continental margin. The breccia contains shocked quartz grains and anomalously high concentrations of Ir. The volume of the deposit is estimated at 250 km^3. The location of the crater is being sought (Leroux et al., 1995; Warme and Sandberg, 1996; Warme and Kuehner, 1998).

Upheaval Dome in Island in the Sky district of Canyonlands National Park, Utah, is also interpreted by some geologists as a meteorite impact crater, with an age possibly ranging from Middle Jurassic to late Tertiary. Yet a recent and exquisitely documented study (Jackson et al., 1998) argued that Upheaval Dome can best be explained as the result of salt tectonics. Salt, derived from the Pennsylvanian Paradox Formation, slowly flowed upward in response to buoyancy forces. Subvertical strata in the center of the dome (central uplift) adjoin the feeder conduit and were pushed upward by flow of the salt. Probably the salt reached the surface before the end of the Pennsylvanian Period, continually breaking through the growing sediment cover. Salt glaciers formed, and during the Middle Jurassic spread into a pancake-shaped glacier some 3 km in diameter. Probably during the time of deposition of the Navajo Sandstone (Jurassic) the feeder stem was pinched off, starving the overlying glacier, which subsequently was eroded away. An equally compelling study by Kriens et al. (1999) presented new data and arguments in favor of an impact hypothesis. Evidence includes shock features such as shatter surfaces and shatter cones, and planar microstructures in quartz grains. The authors argue that the dome formed mainly by centerward motion of rock units along listric faults, as would occur during rebound following collision. The structure is compatible with impact of an asteroid 100–170 m in radius. Thus, the debate over the origin of Upheaval Dome continues. Shaking from a putative meteorite impact has been suggested as the cause of intraformational deformation in the Middle Jurassic Carmel Formation (Alvarez *et al.*, 1998).

Certainly the most significant recent impact event, well preserved in the Southwest and in many areas of the world, occurred 65 Ma at the Cretaceous–Tertiary boundary. It was the event that 'killed the dinosaurs.' Although the impact occurred in the Yucatán region of Mexico, its effect in the Southwest was great, and much of the early evidence of an extraterrestrial origin for the event was developed from outcrops in the Southwest. This impact event is taken up in detail in Chapter 7.

Finally, probably the best known impact crater in the Southwest is Meteor Crater (also known by its original name of Berringer Crater), near Winslow, Arizona. The crater, 1.23 km in diameter and 168 m deep, was formed 50,000 years ago by the impact of an iron meteorite. Although spectacular to view, Meteor Crater was caused by a relatively small object, probably not more than 15–20 m in diameter but traveling at the high speed of some 20 km/s (Schnabel *et al.*, 1999).

The impact record preserved in Phanerozoic rocks of the Southwest is a reminder that the Earth has been struck often by extraterrestrial objects in the past and that impacts by extraterrestrial objects are a rather routine geological process.

4.3 Transition to an active margin: the Antler orogeny

Evidence for active (collisional) tectonics is found in strata of early to middle Paleozoic age in much of the western USA and Canada. In northern California, for example, evidence of subduction as early as Middle Ordovician stems from an assemblage of mafic, clastic, and carbonate strata deformed and metamorphosed under blueschist facies conditions (an indicator of high-pressure, low-temperature metamorphism), and other rocks interpreted to have been deposited in an arc-trench setting, and from the presence of an Ordovician peridotite complex that may have formed in a back-arc basin (Oldow *et al.*, 1989). Middle Ordovician to Silurian tectonism, associated with an arc somewhere offshore, is referred to as the Callahan event (Cotkin, 1992; Cotkin *et al.*, 1992). However, it is now recognized that these rocks are all exotic to North America and were emplaced onto the margin of Laurentia during the middle to late Paleozoic and sub-sequent collisional orogenies. Their formation and possibly their deforma-tion occurred some unknown distance away from the margin. They were swept together and accreted to the margin of Laurentia in a process similar to that in which the Proterozoic crust was formed (Chapter 1).

Not until the Late Devonian and Early Mississippian Periods did the quiescence of the former period end along western Laurentia, and then with thrusting of basinal sediments eastward onto the continental margin (Fig. 4.12) in what is referred to as the Antler orogeny. This orogeny pro-foundly affected the western continental margin, marking the end of the passive margin and the beginning of an active margin that would dominate the tectonics nearly to the present. The Antler orogenic zone is recognized from southern Idaho to the Mojave Desert of southeastern California, but is best expressed in Nevada. The orogen comprises the complex Roberts Mountains allochthon (Fig. 4.12). It consists of lower Paleozoic siliceous sedimentary strata, initially deposited on the continental slope and rise, that were thrust eastward over a contemporaneous shelf and foredeep basin sequence of mainly carbonate rocks. Greenstone incorporated in the thrust wedges is inferred to represent slices of oceanic crust. Once formed, the Roberts Mountain allochthon formed a highland (the Antler Mountains) (Fig. 4.12), imposing a tectonic load that depressed the edge of the continent and formed a deep foreland basin on the remaining un-deformed portion of the miogeocline to the east of the thrust front. Its main feature, the Roberts Mountains thrust, is known from central and northern Nevada. The amount of eastward transport of the allochthon may have been as much as 140 km from its original position (Jansma and Speed, 1995). The allochthon shed debris eastward into the foreland basin (Dickinson, 1981; Ingersoll, 1997) and westward into an extensional basin, the Havallah. The duration of Antler-related events may have spanned about 25 Myr in the Late Devonian and Mississippian. By Pennsylvanian time, the Antler orogen was eroded and below sea level, to be overlapped by shallow-water strata.

Many scenarios have been proposed for the origin of the Antler orogenic belt, but a major difficulty in interpretation has always been the lack of an identifiable arc terrane, specifically the plutonic, volcanic, and volcaniclas-tic rocks, within the orogen. Two models are illustrative. The first model (Dickinson, 1981; Ingersoll, 1997) postulates a *west*-dipping subduction

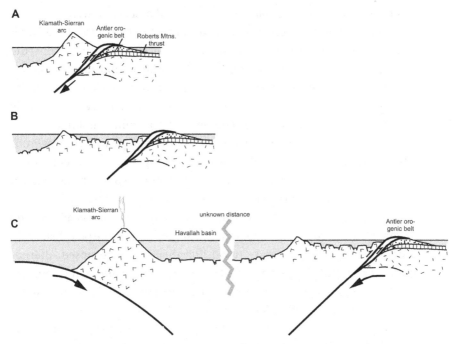

Fig. 4.12. Two models for the driving forces behind the Antler orogeny, both of which invoke west-dipping subduction, are presented in this figure. The first model (A and B) postulates collision of the Klamath–Sierran island arc with the western margin of Laurentia (Burchfiel and Davis, 1972; Dickinson, 1981; Ingersoll, 1997). (A) Hypothetical cross section of the arc and continental margin at time of impact in the Late Devonian to Early Mississippian. Collision of the arc with the continental margin resulted in closing of a forearc basin and thrusting of miogeoclinal sediments onto the western continental margin (forming the Roberts Mountain thrust), and accretion of oceanic crust to the continent. After Burchfiel and Davis (1972). (B) A broad region of extension and subsidence may then have occurred along the collision zone, which could have buried arc volcanic, volcaniclastic, and plutonic rocks from view. Alternatively (not shown), the arc rocks may have been removed by transform faulting parallel to the continental margin (Ingersoll, 1997). The second model (c) does not require collision of the Klamath–Sierran arc with the continental margin. Rather, a second, *east*-dipping subduction zone existed briefly. Sediments of the back-arc basin were thrust onto the continental margin as the margin entered the east-dipping subduction zone. The subduction zone shut down when the buoyancy of the subducting slab became too great.

zone, overlain by the Klamath–Sierran arc and associated trench system, some distance off the western margin of Laurentia. The cause of the Antler event, then, was related to collision of the arc and trench system against the western margin of Laurentia (Fig. 4.12). Slab rollback and consequent migration of the Antler trench and accretionary prism toward the continent resulted in squeezing of the oceanic siliceous sedimentary and volcanic rocks over the carbonate shelf sediments of the continental margin in a great thrust sheet. So, where *are* the arc volcanic and plutonic rocks? Probably some of the arc was eroded away while orogeny was in progress, but this explanation alone does not seem sufficient to explain the absence of arc rocks. Possibly, as the slab continued to roll back toward the continent, extension in the back-arc region expanded eastward and allowed the magmatic rocks to subside and become buried. Yet, despite post-Paleozoic extensional deformation and erosion, none has yet been exposed. Alternatively, transform motion developed along this margin during the Pennsylvanian, and the magmatic rocks were carried away parallel to the margin to an as-yet unrecognized location (Ingersoll, 1997).

The second model (Fig. 4.12), somewhat more complex, postulates separate, oppositely dipping, subduction zones for the Klamath–Sierran arc and Antler thrust belt (Burchfiel and Royden, 1991). The model, which is based on the Appenine system of the Mediterranean Sea, does not require collision of an arc with the continental margin. In contrast with the first model, subduction beneath the Klamath–Sierran arc was *east*-dipping. In response to increased magmatic activity in the Klamath–Sierran arc, a second, *west*-dipping subduction zone was initiated in the Havallah basin behind (east of) the Klamath–Sierran arc. As a result of rollback of the westward-subducting slab, (1) an extensional region, which became larger with time, formed in the Havallah basin between the two subduction zones, and (2) a thrust belt developed in the accretionary prism of the west-dipping subduction zone and migrated toward the continental margin. The eastern thrust belt was jammed onto the margin of Laurentia when the continental margin entered the subduction zone. Thrusting eventually ceased as the buoyancy of the subducting slab became sufficiently great to retard subduction. Arc volcanic rocks that may have been erupted in the hanging wall of the Antler subduction system were within or near a region of back-arc extension (the enlarging Havallah basin) and, consequently, underwent extension and subsidence (thus obscuring them)

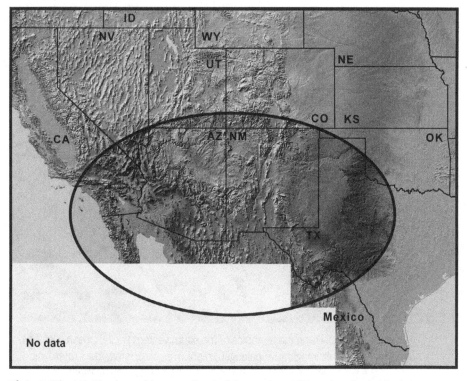

Plate 1 (Fig. I.1). Physiographic map of part of the western USA and northern Mexico compiled from digital elevation data. The Southwest of North America, as loosely defined in this book, is shown in the oval inset. Modified from world-wide web pages by Andrew D. Birrell.

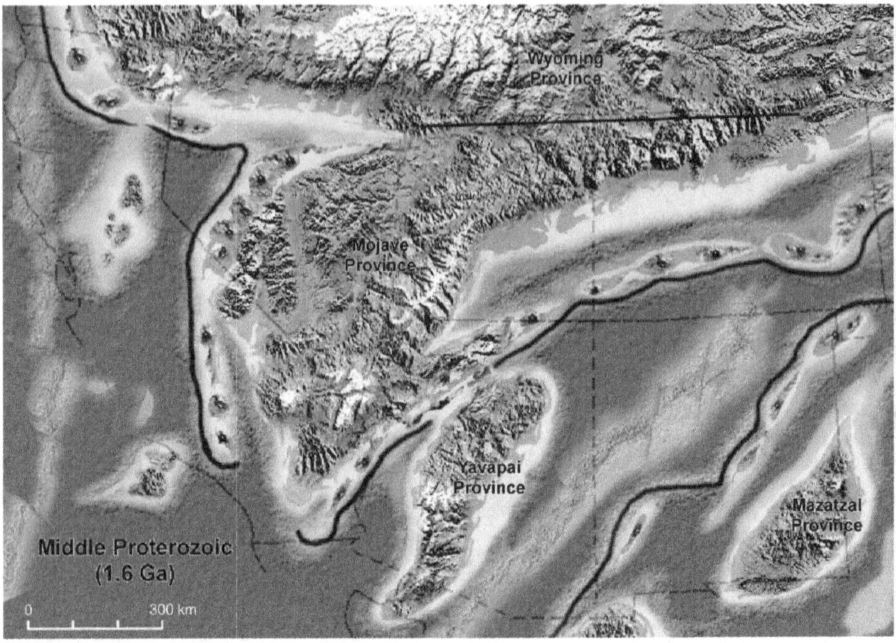
Wyoming
Province
Mojave
Province
Yavapai
Province
Mazatzal
Province
Middle Proterozoic
(1.6 Ga)
0
300 km

Plate 2 (Fig. 1.15). Using the tectonics of the southwestern Pacific Ocean (Fig. 1.14) as an analogue, this imaginative paleogeographic map illustrates how Proterozoic crust in the Southwest may have been created. An array of oceanic islands and island arcs are swept together as they are carried toward their respective subduction zones adjacent to the continental margin. Eventually they collide with, and are accreted to, the continent. This figure illustrates a stage in the development of Laurentia just prior to formation of the Mesoproterozoic Yavapai and Mazatzal terranes of the Southwest. From R. C. Blakey, website: http://vishnu.glg.nau.edu/rcb/paleogeogwus.htm, by permission.

Plate 3 (Fig. 2.3). Mesoproterozoic metasedimentary rocks and diabase sills are exposed in the Salt River Canyon of central Arizona west of US Route 60. Symbols are as follows: uds, the upper member of the Dripping Springs Limestone; lm, am, respectively, the lower and algal members of the Mescal Limestone; tq, the Troy Quartzite. At this locality, the section is inflated about 100% by sills of diabase, which generally form slopes. See Wrucke (1989).

Plate 4 (Fig. 2.5). Red quartzose, silty sandstone of the Proterozoic Dox Formation, comprising the upper part of the Unkar Group (Grand Canyon Supergroup), is exposed along the east side of Tanner Creek in the Grand Canyon. It is separated from the overlying Cambrian Tapeats Sandstone (prominent cliff-forming unit at top of photograph) by an angular unconformity.

Crystal Spring Formation   Beck Spring Dolomite

Plate 5 (Fig. 2.7). The Crystal Spring Formation, seen in this photograph near Saratoga Spring, Death Valley National Park (California), is the lowermost formation of the Mesoproterozoic to Neoproterozoic Pahrump Group. At this exposure, the arkosic lower part of the Formation is not exposed. Massive diabase sills containing septa of stromatolitic carbonate (underlying the lower slopes across the drainage) make up the middle part of the Formation. The upper part consists of alternating beds of carbonate, quartzite, and shale/siltstone.

Plate 6 (Figs. 2.8A,B) (A) Prominent beds of orange-weathering, clay-rich dolostone and intercalated thin quartzite (darker units) of the Beck Spring Dolomite, part of the Mesoproterozoic to Neoproterozoic Pahrump Group. (B) The Beck Spring Dolomite is characterized by stromatolitic dolostone (wavy laminae in upper bed, behind scale) and interbedded quartzite (lower bed). Scale is open 21 cm. Both photographs were taken at the southern end of the Saratoga Hills near Saratoga Spring, Death Valley National Park, California.

Plate 7 (Fig. 5.8). Tilted strata of the Pennsylvanian–Permian Fountain Formation in Garden of the Gods Park, Colorado Springs, Colorado. The Fountain Formation comprises coarse-grained sediments shed eastward off the Front Range of the Ancestral Rocky Mountains into the Denver Basin. Uplift of the strata to their current, near-vertical position occurred during the much later Laramide orogeny.

Plate 8 (Figs. 5.12A,B) Continental red beds are characteristic of the Permian Period in much of the Southwest and are especially prominent on the Colorado Plateau. (A) These colorful strata exposed in Oak Creek Canyon, Arizona, comprise the Hermit and Schnebly Hill Formations (Blakey, 1990b), deposited relatively far from the Uncompahgre uplift. Fm, Formation. (B) The brick-red Halgaito Shale, a lateral and finer-grained equivalent of the Cutler Formation, forms the lower, slope-forming portion of monuments in the Valley of the Gods, southern Utah. The upper, cliff-forming unit is Cedar Mesa Sandstone.

Plate 9 (Fig. 5.12E). Sandstone of the Cutler Formation forms the lower two-thirds of this outcrop, exposed along US Highway 84 seven kilometers west-northwest of Abiquiu, New Mexico. The overlying brownish sandstone beds are the lowermost part of the Triassic Chinle Formation.

Plate 10 (Fig. 5.15). The DeChelly Sandstone, seen in this photograph from near Round Rock, Arizona, is a dune deposit formed as winds scoured alluvial flats draining from the Uncompahgre Mountains.

Plate 11 (Fig. 5.21). The magnificent El Capitan in the southern Guadalupe Mountains, Texas, viewed from the east, exposes forereef talus beds dipping toward the south (left). Siltstone and sandstone of the Permian basin (Brushy Canyon Formation) underlie the slopes beneath the reef limestone.

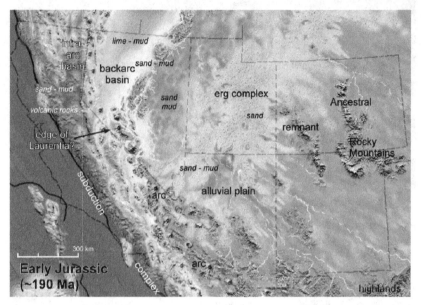

Plate 12 (Fig. 6.10). Paleogeographic map of part of the western USA during the Early Jurassic Period (approximately 190 Ma). The Southwest lay in the tradewind belt. Winds blowing persistently from the northeast kept the southwestern margin of the continent as dry as the modern Sahara Desert, and sand seas occupied much of the region. Modified from R. C. Blakey, website: http://vishnu.glg.nau.edu/rcb/paleogeogwus.htm, by permission.

Plate 13 (Fig. 6.12). At Valley of Fire State Park, Nevada, the contact between fluvial sandstone, siltstone, and mudstone of the Moenave–Kayenta Formation (undifferentiated) and eolian dune deposits of the overlying Aztec Sandstone is transitional. Seen in this photograph near the Park Visitor Center, cross-bedded sandstone layers (white) are interbedded in the upper Moenave–Kayenta Formation (Marzolf, 1988).

Plate 14 (Fig. 6.17). Part of the Jurassic section at Red Mesa, north of Mexican Water, Arizona. At this location the lower part of the Summerville Formation is characterized by contorted strata. Stratigraphic identification from S. Lucas and S. Sempken (personal communication).

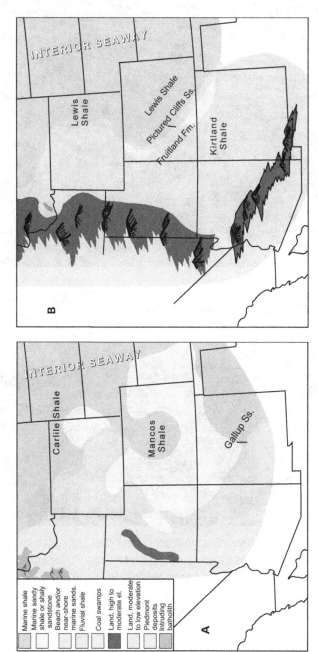

Plate 15 (Fig. 7.4). Paleogeographic maps of the western margin of the Cretaceous Western Interior seaway of Late Cretaceous time. Selected characteristic formations are identified. The relationships of different lithologies to sea level, and the changing environments of deposition through time, are indicated. (A) The seaway approximately 90 Ma, during a minor regression in the Mancos sea. (B) Approximately 74 Ma, during the final withdrawal of the Lewis sea. Times are indicated in Fig. 7.5. From McGookey *et al.* (1972).

A

B

Plate 16 (Fig. 7.14A,B). The Naashoibito Member of the Kirtland Formation contains numerous petrified logs, such as the one shown in this photograph. The location is the De-Na-Zin Wilderness, south of Farmington, New Mexico. Scale is in cm.
(B) Details of knots and even of the grain structure of this 74 m.y.-old tree in the Kirtland Formation are well preserved. Location is same as (A); scale is in cm.

Plate 17 (Fig. 7.15). Cretaceous–Tertiary boundary layers are preserved in coal-bearing sediments of the Raton Formation in northern New Mexico and southern Colorado. The prominent white layer is the claystone layer, composed of altered glassy ejecta launched ballistically from the Chicxulub impact site. It represents a bad day for the Earth. Above it (immediately below the scale) is the impact (or 'fireball') layer, formed as shocked rocks from the impact area, lofted into the atmosphere by a fireball, settled out over the subsequent weeks and months. This layer contains grains of shocked quartz and feldspar, and high concentrations of iridium. Deformation of the claystone and impact layers is due to the weight of the overlying strata. The boundary layers overlie a layer of carbonaceous shale and are overlain by a bed of coal. See Box 7.3. Scale is in cm. Location is 4.6 km south of Starkville, Colorado. Reference: Pilmore and Flores (1987); Izett (1990).

Plate 18 (Fig. 7.17). The Raplee monocline, near Mexican Hat, Utah, resulted from crustal shortening during the Cretaceous–Tertiary Laramide orogeny.

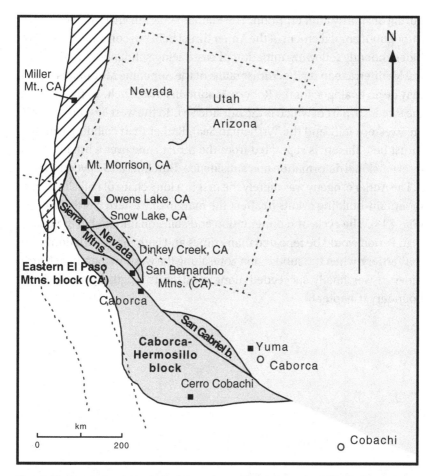

Fig. 4.13. Reconstructed middle Paleozoic continental margin of the Southwest. Light gray is more-or-less stable continent. Medium gray signifies parts of the middle Paleozoic continental margin displaced in post-middle Paleozoic time. Ruled pattern indicates Antler-belt rocks displaced in post-middle Paleozoic time. Solid lines are present state lines on the stable continent. Dashed lines are boundaries relative to the Caborca–Hermosilla block in its present location to emphasize the displacement of this terrane. Squares are cities and towns restored to their post-Antler locations. Displacement of this margin may have occurred coincident with the Sonoma orogeny of Late Permian to Triassic time (Chapter 5). CA, California; b., block; Mtns., Mountains. From Stevens *et al.* (1992).

during or shortly after eruption. The Klamath–Sierran arc remained active throughout emplacement of the Antler thrust belt, but continued to evolve independently following initiation of east-facing subduction. In the central Mediterranean region, thrust faults of the Appenine Mountains (Italy) may be an analogue for the Roberts Mountains thrust belt. To the east lies the foredeep, part of which is the Adriatic Sea. To the west lies the volcanic arc (western Italy and the Tyrrhenian Sea). Rather than colliding with the thrust belt, the arc is separated from the thrust zone over a broad region by extensional deformation and subsidence (Burchfiel and Royden, 1991).

The Antler orogeny was merely the first in a long chain of collisional and mountain-building events to affect the margin of western North America (Fig. 4.13). The cycle of sedimentation and collision initiated in the Devonian Period would be repeated many times and with many variations into the Tertiary, when the subduction zone along the western margin of North America was finally succeeded along most of its length by a transform boundary (Chapter 8).

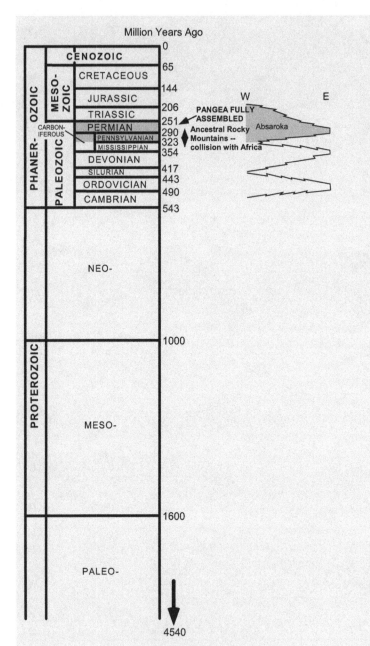

Geologic time scale. For explanation see p. 3.

5

Pangea! Collision with Gondwana

Pennsylvanian and Permian

The famous Mittens of Monument Valley in northern Arizona are part of the dramatic and colorful landscape created by Permian rocks over wide areas of the Colorado Plateau. The valley floor and sloping bases of the monuments are cut in the Organ Rock Shale, a lateral equivalent of the Permian Cutler Formation. The vertical cliffs are the eolian Permian DeChelly Sandstone. The capping rocks are Triassic Moenkopi Shale and Shinarump Conglomerate (Chinle Formation).

5.1 Introduction

During the late Paleozoic, the geological development of the Southwest was controlled by two major events. The first of these, one of the major global events of the Phanerozoic era, was the final assembly of the supercontinent Pangea, which profoundly affected the Southwest. In fact, the Southwest found itself in a great squeeze (Fig. 5.1). From the southeast came Gondwana, which when crunched together with Laurentia, completed the assembly of Pangea. From the west, riding on oceanic crust, came bits and pieces of island arcs and perhaps a microcontinent, to be accreted to the western margin of Pangea when the supercontinent entered the subduction zone. In addition, near the beginning of the Pennsylvanian Period, global sea level rose again, marking the Absaroka transgression. The wide range in topography that resulted from continental collision, coupled with the changing sea level as marine water inundated then retreated from the continental margin, created a wide range of paleogeographic environments over relatively short distances.

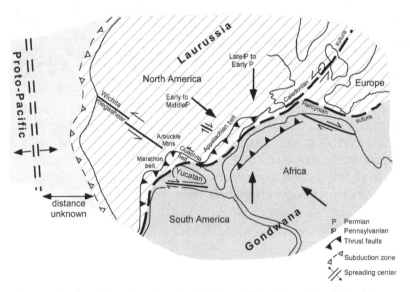

Fig. 5.1. Configuration of continents at the time of late Paleozoic collision between Laurussia (Laurentia and Russia) and Gondwana, and formation of the Ouachita–Marathon orogen and the Ancestral Rocky Mountains. Relative plate motion is shown by heavy arrows. 'Mtns' is 'Mountains'. From Budnik (1986).

5.2 Ancestors of the Rockies

A profound tectonic event or combination of events, affecting the eastern and southern margins of North America from New England to the Yucatán region and much of the western interior of the United States, began in the Late Mississippian, reached its greatest intensity in the Middle Pennsylvanian, and ended in the Early Permian. In the West and Southwest, deformation resulted in thrust belts and basins in central and west Texas (the Ouachita–Marathon thrust belt) and in a series of high mountain ranges and deep, bounding basins that stretched from Oklahoma across the Texas Panhandle and parts of New Mexico and Colorado to central Utah (Fig. 5.2). These ranges are referred to as the Ancestral Rocky Mountains.

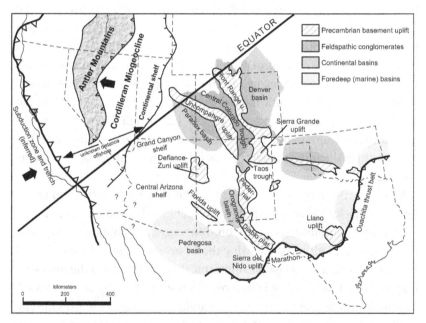

Fig. 5.2. Tectonic setting of the Southwest during the late Paleozoic Era. Uplifts and basins along the plate margins are associated with the Cordilleran miogeocline and Ouachita–Marathon fold and thrust belt, and on the craton reflect the Ancestral Rocky Mountains orogen. U., uplift; plat., platform. The timing of different tectonic elements is in part diachronous. Simplified from Handschy and Dyer (1987), Blakey and Knepp (1989), and Soegaard (1990).

Two competing hypotheses have been advanced to explain this widespread deformational event. The first, widely accepted, relates deformation throughout the western and southwestern United States to collision with Gondwana, the southern continent. As the proto-Atlantic Ocean closed, northwestward movement of Gondwana and eastward drift of Laurentia brought Africa into a crushing collision with eastern North America (Fig. 5.1). West- to northwest-directed convergence between Gondwana and Laurentia occurred mainly along the Appalachian zone of eastern North America, resulting in hundreds of kilometers of crustal shortening (Budnik, 1986). Along the southern and southeastern continental margin, right-oblique transpression resulted in formation of the Ouachita orogen, which extended westward to the Big Bend region of Texas, and thence southward into Coahuila. When Gondwana pulled away from Laurentia in the Triassic (next chapter), it left a piece behind: the Coahuila terrane (Handschy *et al.*, 1987). Deep within the plate, in southwestern North America, crustal shortening was largely absorbed by oblique slip along an ancient zone of weakness, which was critically located approximately on strike with the eastern Ouachita belt and at an offset in the southern margin of North America. Referred to as the Wichita megashear and stretching from Oklahoma to Utah, this zone underwent 120–150 km of left lateral strike slip and crustal compression. Stresses created by transpression along the Wichita megashear resulted in ranges of the Ancestral Rocky Mountains (Fig. 5.2).

An alternative explanation (Ye *et al.*, 1996) separates formation of the Ancestral Rocky Mountains from the Ouachita–Marathon orogen, based on the style of deformation and on differences in timing and direction of compression. Although allowing the Ouachita orogen to have formed by collision of Laurentia with Gondwana, this model relates the Ancestral Rocky Mountains (southern Oklahoma to central Utah) to northeastward subduction along the southwestern margin of Laurentia (off present-day northern Mexico). Evidence for a subduction zone is inferred from the presence of a volcanic arc in east-central Mexico and from related coeval deformation, and is presumed to have been continuous with the subduction boundary known to have persisted along the western margin of Laurentia beginning in the Devonian, as discussed in the previous chapter.

Whereas farther north the arc lay offshore and was separated from Laurentia by oceanic crust, the volcanic arc of east-central Mexico lay directly on
the continent. Based on the structure of the subduction zone beneath the
Sierra Pampeanas in the eastern Argentinian Andes (further discussed in
Chapter 7), geologists infer that the subducted slab dipped very shallowly
beneath Laurentia, allowing it to couple with – and deform – the overlying
plate far inland. In this model, then, the Ancestral Rocky Mountains are
attributed to intraplate deformation within the overriding plate (Laurentia) of the subduction boundary (Ye *et al.*, 1996). One problem with this
model is the postulated northeastward dip of the subduction zone, which
contrasts with the inferred westward dip of subduction during the Antler
(Mississippian) (Chapter 4) and the Sonoma (Late Permian to Triassic)
(Chapter 6) orogenies.

Thus, no general consensus on the origin of the Ancestral Rocky Mountains exists, and the origin of the Ancestral Rockies will probably remain
a topic of research for many years. Whatever the cause, the Ancestral
Rocky Mountains orogeny created a very complex paleogeography that
dominated sedimentation patterns for the next approximately 100 Myr
(through the Triassic). Among the great mountain ranges created during the Ancestral Rocky Mountains orogeny was the Uncompahgre uplift,
which stretched from northern New Mexico to central Utah (Fig. 5.2). As
is discussed in this and the next chapter, many of the prominent Permian
and Triassic red bed units that crop out over wide areas of the Colorado
Plateau were shed from the Uncompahgre Mountains. These units were
largely alluvial fans and the more distal equivalent units. Along the southwest side of the Uncompahgre range lay a deep, narrow basin, the Paradox
basin. Over 6.7 km of sandstone, conglomerate, evaporite minerals, and
limestone were deposited in this basin. Other major mountain ranges –
the Front Range, Pedernal, and Diablo – stretched from central Colorado
through New Mexico to Texas, while the Defiance–Zuni and Florida uplifts dominated western New Mexico and eastern Arizona. Between the
Front Range and Uncompahgre uplifts lay the Central Colorado trough,
which stretched into northern New Mexico as the Taos trough (Fig. 5.2).
Although these ranges are long gone and the basins filled, evidence for
their existence is preserved in the patterns of sedimentation throughout
the Southwest.

5.3 Paleogeography of the Pennsylvanian

Beginning in the Early Pennsylvanian, the seas once again flooded the continent, and by the Middle and Late Pennsylvanian, open-marine waters covered large parts of the Southwest. This marine highstand marked the Absaroka transgression. The transgression was not smooth, however. Sea level changed constantly, withdrawing and advancing differently in different areas in response to ongoing deformation of the Ancestral Rocky Mountains. Ranges of the Ancestral Rockies poked up as rugged islands in the seas (Fig. 5.3). Thus, the strata that characterize the Pennsylvanian differ widely in lithologic character over relatively short lateral and vertical distances, depending on their proximity to the ranges of the Ancestral Rockies and on the shifting seas.

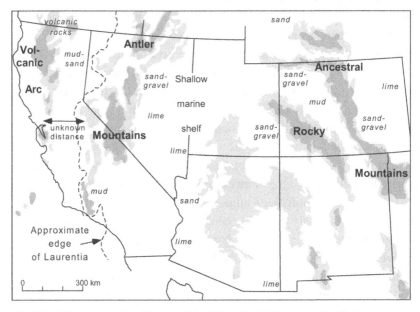

Fig. 5.3. Paleogeography of the western USA in Early Pennsylvanian time (*c.* 315 Ma). Uplifts of the Ancestral Rocky Mountains poke up through shallow marine waters of the Absaroka transgression. Local environments of deposition range widely because of the complex paleogeography. Shaded areas are subaerial, with darker shading indicating highlands. Compare Fig. 5.2. Modified from R. C. Blakey, website http://vishnu.glg.nau.edu/rcb/paleogeogwus.htm.

Over large parts of the Southwest, Pennsylvanian seas spread over a karst topography, characterized by caverns, underground passages, and sinkholes, developed on Mississippian limestone. Long exposure to subaerial weathering created a deep, red soil over much of the region. Under the advance of the Pennsylvanian seas, the soil was reworked to form a widespread and thin reddish to purplish mudstone, which marks the Mississippian to Pennsylvanian unconformity (Molas and Log Springs Formations).

Marine waters invaded the Southwest from two directions. From the south, waters spread out of the Pedregosa basin onto the central Arizona shelf of southeastern Arizona, New Mexico, and northern Mexico. From the northwest and north, the seas spread onto the Grand Canyon shelf from the Cordilleran miogeocline (Fig. 5.2). In many areas, initial Pennsylvanian deposits consisted of sandstone and/or conglomerate, in places containing plant fragments. These deposits represent beach and near-shore deposits formed as seas transgressed onto the continent. Along continental margins adjacent to positive areas, tidal flats and deltas developed. Sandstone, mudstone, and shaly limestone, commonly with plant fragments and fossil wood, interfingered toward deeper water with platform carbonate. Terrigenous influx occurred across the shelf at various times and from different source areas, reflecting local tectonic disturbances (Blakey and Knepp, 1989; Wilson, 1989).

In these marine basins and platform areas, Pennsylvanian strata are dominantly carbonate, typically fossiliferous. A plethora of names exists for these strata (Fig. 5.4), for two reasons. First, carbonate rocks in different parts of the Southwest do not exactly correlate because the timing of tectonic events was diachronous in the different regions. Typically, broad regions of shallowly submerged craton were separated by intervening positive areas (Figs. 5.2, 5.3) which became elevated at slightly different times (Fig. 5.2). Secondly, in some areas such as central and southern New Mexico, which were essentially all part of a single platform, strata are exposed primarily in isolated mountain ranges widely spread throughout the region. Therefore the description and correlation of these strata have been rather piecemeal. They have been recognized, described, and named by different geologists working over a span of more than 90 years. Equivalent strata in adjacent mountain ranges may have different names (Myers, 1982; Wilson, 1989). Fortunately, the presence of abundant and distinctive

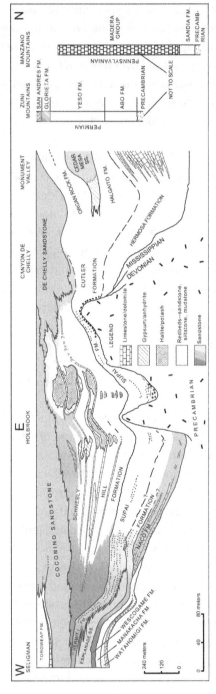

Fig. 5.4. Restored section of Pennsylvanian and Permian rocks across part of the Colorado Plateau from western Arizona eastward, then northward to northeastern Arizona (section is not straight). Thicknesses are shown relative to the top of the Permian. This section shows the great complexity of rock types derived from the complex paleogeography of the time, and the resulting plethora of formation names for rocks of equivalent age. For comparison, stratigraphic columns of generally equivalent rocks in western and central New Mexico are shown (not to same scale). FM., Formation; SS., Sandstone. Cross section from Blakey (1990b); New Mexico columns from Colpitts (1989) and Myers (1982).

Fig. 5.5. Nodular layers of gray limestone 5–20 cm thick alternate with layers of silty shale in the middle part of the La Pasada Formation, seen in this outcrop in Dalton Canyon 9.3 km north of Pecos, New Mexico. The La Pasada Formation of the southern Sangre de Cristo Mountains is approximately equivalent in age to the lower part of the Madera Group (Kues, 2001), but contains much more sand and silt because of its close proximity to uplifts of the Ancestral Rocky Mountains. Scale is 1 m.

fossils makes correlation possible, after enough work. Examples of these widespread platform carbonate rocks are the Madera Group (Fig. 5.5), which prominently caps the Sandia Mountains above Albuquerque, New Mexico. Equivalent rocks (with different names) are exposed in mountain ranges southward into Mexico. In the Grand Canyon, Pennsylvanian carbonates are present but are slightly older (Watahomigi Formation). Here, strata equivalent in age to the Madera Group consist mainly of sandstone and mudstone (Blakey, 1990a, b) (Fig. 5.4). These differences reflect diachronous events on continental shelves separated from each other by a persistent broad upwarp (Fig. 5.3).

In basins adjacent to the ranges of the Ancestral Rockies, debris fans accumulated. In central and northern New Mexico, for example, Pennsylvanian strata consist of fine- to coarse-grained siliciclastic rocks deposited adjacent to the Pedernal uplift in central New Mexico and in the Taos trough, a southern portion of the central Colorado trough between

A

B

Fig. 5.6. (A) These interbedded mudstone and sandstone layers, part of the Sangre de Cristo Formation, were deposited on a floodplain adjacent to an active, braided stream or river system draining from the north. Sandstone layers may represent flood events that spilled sand out of river channels onto a muddy floodplain (Soegaard and Caldwell, 1990). Scale is one meter. (B) Convoluted bedding in fine-grained sandstone and mudstone of the upper Sangre de Cristo Formation is interpreted as slump folds formed in semiconsolidated sediments deposited on an oversteepened slope. Deformed beds pass vertically into undeformed strata above and below. Both exposures are in the Sangre de Cristo Mountains east of Peñasco, New Mexico.

A

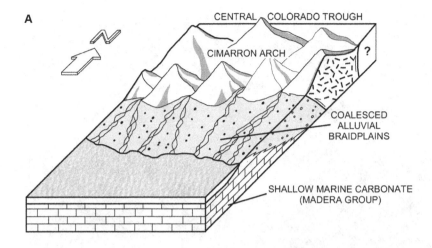

B

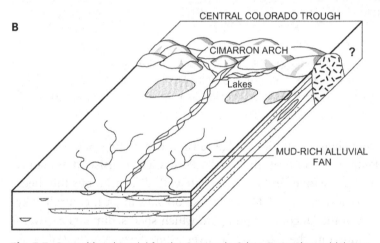

Fig. 5.7. Depositional model for the Sangre de Cristo Formation, which was deposited in the active Taos trough adjacent to the Uncompahgre and Pedernal uplifts. (A) The lower part of the formation (lower megasequence) consists of gravel and sand deposited on coalesced braidplains adjacent to the mountain front. (B) The upper part of the formation comprises sand and mud deposits of a braided river channel which migrated across a broad, muddy floodplain of low relief. Overall the Sangre de Cristo Formation apparently records a lowering of topographic relief in the source area to the north. Modified from Soegaard and Caldwell (1990).

Fig. 5.8. Tilted strata of the Pennsylvanian–Permian Fountain Formation in Garden of the Gods Park, Colorado Springs, Colorado. The Fountain Formation comprises coarse-grained sediments shed eastward off the Front Range of the Ancestral Rocky Mountains into the Denver Basin. Uplift of the strata to their current, near-vertical position occurred during the much later Laramide orogeny. See also Plate 7.

the Uncompahgre uplift and the Sierra Grande arch (Fig. 5.2). Here, the lowest Pennsylvanian unit, the Sandia Formation, consists dominantly of siltstone, sandstone, and conglomerate with discontinuous thin beds of marine limestone (Fig. 5.6). Its deposition was, in part, controlled by early uplift of the Uncompahgre uplift, which shed sediments eastward and southward into the Taos trough (Myers, 1982; Soegaard and Caldwell, 1990). The Sandia Formation represents a clastic wedge of coarse-grained, braided-alluvial sediments deposited in the Taos trough in response to renewed uplift of the Uncompahgre uplift (Soegaard and Caldwell, 1990).

The Sandia Formation is separated by southward-thickening shelf carbonates of the La Pasada and Alamitos Formations (which are approximately laterally equivalent to the Madera Group farther south in New Mexico) from a second siliciclastic unit, the Sangre de Cristo Formation. The Sangre de Cristo Formation, which crops out throughout the southern Sangre de Cristo Range, comprises a thick package of coarse-grained sandstone, conglomerate, and mudstone (Fig. 5.6). Rocks display

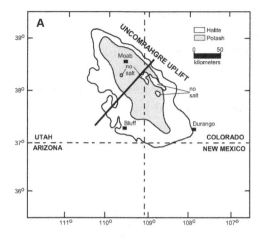

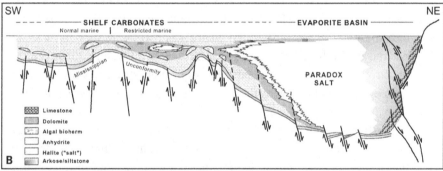

Fig. 5.9. (A) Map showing location of the Paradox basin and of salt facies. Gray area is extent of high-grade potash in the basin. Salt is missing from synclinal troughs owing to flow into adjacent anticlines. The heavy line is the location of cross section shown in (B). From Williams-Stroud (1994). (B) Schematic cross section across the Paradox basin palinspastically reconstructed at late Middle Pennsylvanian time. From Stevenson and Baars (1987).

very complicated trough and planar crossbedding, convolute bedding, and other sedimentary features indicative of deposition in a subaerial alluvial environment. A lower 'megasequence' is indicative of deposition as a network of shifting, shallow channels on a broad outwash plain, reflecting numerous coalescing braided streams. An upper megasequence represents deposits of an active braided river migrating across a broad, muddy floodplain (Fig. 5.7) (Soegaard and Caldwell, 1990). Southward, the Sangre de Cristo Formation grades into limestone of the Madera Group deposited on the continental shelf of southern New Mexico.

Similarly, the Fountain Formation, spectacularly exposed along the modern Front Range of central and northern Colorado (Fig. 5.8), comprises coarse-grained sediments shed off the eastern side of the Pennsylvanian–Permian Front Range uplift into the Denver basin (Fig. 5.2). Originally laid down on alluvial fans and plains, these sedimentary strata were subsequently raised into their present inclined position by later uplift (Chapter 7).

5.4 The Paradox basin

Probably the best studied of the basins associated with the Ancestral Rockies is the 320-km long Paradox basin (Fig. 5.9), which lies adjacent to the Uncompahgre uplift. The basin is asymmetrical, with the deepest part next to the uplift. Because of the oil and gas that lie trapped in rocks deposited in the basin, the Paradox basin has been well studied, primarily through drilling. As a result of complicated structures and unusual lithologies associated with the basin, it has proven paradoxical in more than just name.

Sub-Pennsylvanian strata are depressed more than 6 km below the surface but, of course, the basin was formed gradually so that water depth in the basin was never great. Sedimentary strata deposited in the Paradox basin comprise the Middle Pennsylvanian Paradox Formation, 1800–2100 m of evaporite beds (Fig. 5.9). Altogether, 33 cycles of salt were deposited. Each cycle consists of thick halite beds separated by interbeds containing anhydrite, silty dolostone, and black carbonaceous shale. The halite beds range from 7 to 270 m thick in the deepest part of the basin, pinching out on the flanks of the basin. More than half of the halite units contain primary sylvite (KCl). The interbeds average about 10 m in thickness.

A 'paradox' associated with the evaporite deposits of the Paradox Formation is the paucity of magnesium sulfate salts (Williams-Stroud, 1994). Modern seawater (the model for Pennsylvanian seawater) contains a relatively high abundance of magnesium relative to calcium, and would be expected to precipitate magnesium salts in abundance before potassium salts. Yet, sylvite is a typical constituent of the Paradox Formation, whereas magnesium sulfate salts are not. Obviously, the evaporite deposits of the Paradox Formation cannot be explained by equilibrium evaporation of

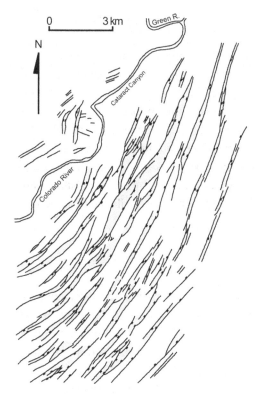

0   3 km

N

Fig. 5.10. Simplified map of graben systems ('The Grabens') in the Needles district of Canyonlands National Park, Utah. The complex structures result from gravitational gliding of a brittle plate of sandstone and shale 450–500 m thick over ductile evaporite strata of the Hermosa Formation (Paradox Member) as downcutting by the Colorado River has released the northwestern lateral buttress of the plate. Simplified from Trudgill and Cartwright (1994).

modern seawater. A likely scenario for deposition of the salts is that individual cycles resulted from fluctuations in sea level. Periodic rises in sea level flooded the basin, establishing open communication across a shallow sill with the open ocean. After the basin was cut off from seawater inflow, a perennial saline lake was established. Evaporation concentrated the sea water, forcing different salts to precipitate as their saturation concentrations were reached. However, the composition of the marine water in the basin was altered by exchange of Mg in the water with Ca in the Mississippian limestones in marginal parts of the basin, producing a calcium-enriched brine (Williams-Stroud, 1994). The result was to deplete the water in Mg and suppress the precipitation of magnesium salts. Enough potassium salts are present in the Paradox basin to constitute an economic resource. This cycle must have been repeated 33 times. Smaller-scale layering in the calcium sulfate and halite beds was the result of storm-driven influxes of seawater, which overtopped the sill. The storm cycles are inferred to have had an average periodicity of several years.

Finally, salt flows. Because salt is less dense than most other rocks, it is buoyant and tends to rise through the overlying rocks, penetrating and disrupting their stratification. Flowage of salt from the Paradox Formation makes it difficult to determine the original thickness of the salt with certainty. The upward movement of salt with the consequent collapse of strata overlying the salt beds has given rise to the fabulous structure of the Needles district of Canyonlands National Park (Fig. 5.10). Diapirs of salts – salt walls – which intruded along fault zones underlie the Paradox, Salt, Moab, and Lisbon Valleys, among others, of Utah, forming collapsed anticlines.

5.5 Permian red beds

The Pennsylvanian Period gave way to the Permian with little change in tectonic conditions, save for gradually declining sea level. The continental seas of the Absaroka transgression slowly withdrew, leaving continental alluvial sediments to replace marine sediments. Marine sediments were followed by widespread alluvial fan deposits, derived mainly from the Uncompahgre and other uplifts, or by more distal continental sediments. Uplifts of the Ancestral Rockies remained high and a major source for debris, but were rapidly eroded. Overall, the paleogeography of the Southwest remained diverse.

An arid climate led to formation of red beds across wide portions of Pangea. Because Pangea was subsequently fragmented, red beds of Permian age now occur in parts of Europe, Asia, Africa, and South America. Permian red beds are very characteristic of the Southwest, particularly of the Colorado Plateau, where they are the most widespread geologic period exposed.

The Uncompahgre Mountains, which overlooked a coastal plain to the west and south, continued to dominate sedimentation. Huge alluvial debris fans were formed at the foot of the mountains in a belt 50–65 km wide stretching along the range from eastern Utah to northern New Mexico. Here, the sediments consist of a wedge of coarse-grained arkose, the Cutler Formation, up to 3700 m thick east of Moab, Utah, near the mountain front (Fig. 5.9). Sediments were progressively finer-grained farther from

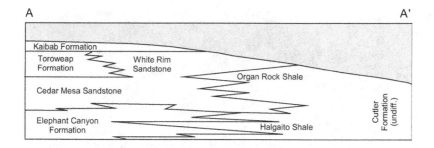

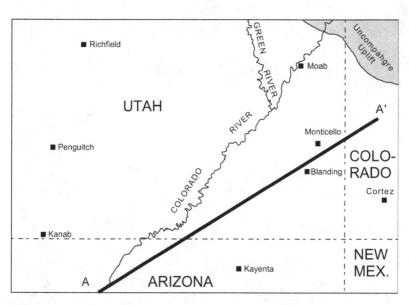

Fig. 5.11. The Uncompahgre uplift dominated sedimentation throughout large areas of the Southwest during the Permian Period. In this figure (upper panel) the correlation of some prominent Permian units in the Four Corners area of the Colorado Plateau is shown. The approximate line of cross section A–A' is indicated on the map (lower panel). The Permian Period was a time of great variability across the Southwest, with sedimentation dominated by the Uncompahgre and other uplifts of the Ancestral Rocky Mountains. Closest to the Uncompahgre Mountains, thick deposits of coarse-grained arkose – the Cutler Formation – were formed on alluvial fans built against the mountain front. On alluvial plains of low topographic relief farther from the mountains, finer-grained sand, silt, and mud, which can be distinguished into a number of formations, were deposited. Modified from Baars (1983, 1988).

Fig. 5.12. For caption see p. 136.

the mountain front as the topographic relief diminished, such that only mud and fine sand were deposited on the coastal lowlands to the south and west. The fine-grained sediments are distinguished into several formations, depending on lithology and location. In southern Utah and northern Arizona, they form the Halgaito Shale, Cedar Mesa Sandstone, Organ Rock Shale, and DeChelly Sandstone, all grouped together as part of the Cutler 'Group' (Fig. 5.11). See also Plate 8A,B.

Fig. 5.12. (*continued*)

Fig. 5.12. Continental red beds are characteristic of the Permian Period in much of the Southwest and are especially prominent on the Colorado Plateau. (A) See also Plate 8. These colorful strata exposed in Oak Creek Canyon, Arizona, comprise the Hermit and Schnebly Hill Formations (Blakey, 1990b), deposited relatively far from the Uncompahgre uplift. Fm, Formation. (B) See also Plate 8. The brick-red Halgaito Shale, a lateral and finer-grained equivalent of the Cutler Formation, forms the lower, slope-forming portion of monuments in the Valley of the Gods, southern Utah. The upper, cliff-forming unit is Cedar Mesa Sandstone. (C) In this detailed photo of the Halgaito Shale, channel cut-and-fill structure, conglomeratic lenses, and reduced (green) layers (possibly originally organic-rich?) are all characteristic of deposition in an alluvial environment. Height of outcrop is *c.* 2 m. Photo from Valley of the Gods. (D) The unit forming these slopes in Monument Valley, Arizona, is the Organ Rock Shale, another lateral equivalent of the Cutler Formation. This unit forms the lower slopes of the Monuments throughout the valley. The upper cliffs are eolian DeChelly sandstone. (E) Sandstone of the Cutler Formation forms the lower two-thirds of this outcrop, exposed along US Highway 84 seven kilometers west-northwest of Abiquiu, New Mexico. The overlying brownish sandstone beds are the lowermost part of the Triassic Chinle Formation. See also Plate 9.

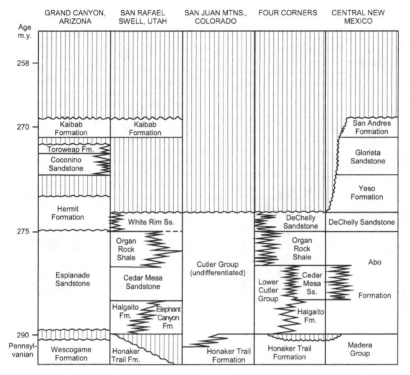

Fig. 5.13. Correlation of Permian units across parts of the Colorado Plateau. Fm., Formation; Ss., Sandstone. Modified from Baars (1974) and Blakey (1990a). Ages from Haq and Van Eysinga (1987).

The lithologic character of the sediments was highly varied, depending on the local depositional environment (Figs. 5.11, 5.12, 5.13). The units interfinger in complex relationships, reflecting the varied paleogeography of western Pangea during the Permian. One such unit, the Cedar Mesa Sandstone, is illustrative of the complexity of Cutler sedimentation. Despite the title of this section, the Cedar Mesa is not a red bed, but rather a white, mainly eolian sandstone comprising dune, interdune, and sand-sheet facies, part of a widespread **erg** (see Box 6.1) that formed on the alluvial plain. To the east, in the Needles District of Canyonlands National Park, the nearly pure white, crossbedded strata of the Cedar Mesa Sandstone are interbedded in a dramatic transition with red, coarse-grained fluvial sandstone of the undifferentiated Cutler Formation (Fig. 5.14). To the northwest, the Cedar Mesa Sandstone interfingers with marine carbonate

A

B

Fig. 5.14. Strong evidence for a dune origin for the Cedar Mesa Sandstone is provided by sweeping crossbeds (A), exposed near the base of Sipapu Natural Bridge, Natural Bridges National Monument, Utah. Here the Cedar Mesa is more than 360 m thick. Only 70 km to the north in the Needles District of Canyonlands National Park, the Cedar Mesa Sandstone (white) forms thin interbeds, which interfinger with red sandstone and mudstone of the Cutler Formation (B).

Fig. 5.15. The DeChelly Sandstone, seen in this photograph from near Round Rock, Arizona, is a dune deposit formed as winds scoured alluvial flats draining from the Uncompahgre Mountains. See also Plate 10.

strata of the Elephant Canyon Formation (or 'lower Cutler beds' of Condon, 1997). To the southeast, they interfinger with evaporite deposits and limestone, lumped as part of the Cedar Mesa. Interpretation of the paleogeography suggests that the Cedar Mesa Sandstone represents a dune field located along the western coastline of Pangea. The sand, originating both from a marine environment to the northwest and from a fluvial environment to the northeast, was deposited as dunes that migrated toward the southeast and terminated in a coastal sabkha (Stanesco and Campbell, 1989). In contrast with thin layers displayed in Canyonlands, the Cedar Mesa is 366 m thick at Natural Bridges National Monument, just 60 km to the south. Eventually, the dune fields comprising the Cedar Mesa Sandstone were buried beneath the deposits of streams and rivers that wound across the flood plain adjacent to the western shoreline, near the present Utah – Nevada border. These deposits – mud, silt, and sand – were the distal equivalent of the coarse-grained Cutler, laid down closer to the front of the Uncompahgre Mountains. They now form the Organ Rock Shale, another classic red bed formation of southeastern Utah (Fig. 5.12D). As the area dominated by dune sands expanded, winds blowing southward across the alluvial sand flats draining from the Uncompahgre Mountains picked up sand and dust, building thick dune deposits in northern Arizona, southern Utah, and northwestern New Mexico. These deposits, which exceed 300 m in thickness, make up the DeChelly Sandstone, well exposed in Monument Valley (chapter photo) and in the type area of Canyon DeChelly (Fig. 5.15). Collectively, these rocks, which now dominate much of the Permian strata in the Four Corners region, make up the Cutler Group. To the south, a blanket of cross-bedded sand spread across northern and central Arizona as another major erg deposit formed on the continental margin. This deposit formed the present Coconino Sandstone, well exposed in the Grand Canyon (Middleton *et al.*, 1990). The Coconino grades eastward into New Mexico as the Glorieta Sandstone. Although not all exactly contemporaneous, the eolian deposits of the Cutler Group – the Cedar Mesa, White Rim, and DeChelly – and the Coconino together make up a very extensive erg deposit that stretched as far north as Montana.

Once again marine waters spread onto the western Colorado Plateau. A relatively minor transgression and the ensuing regression are recorded

Fig. 5.16. The prominently cross-bedded White Rim Sandstone forms a thin but very resistant unit over wide areas of southeastern Utah. Removal by erosion of overlying strata exposes the Sandstone, creating the laterally extensive 'White Rim.' The Sandstone is also visible on the opposite side of the canyon, where it appears as a narrow band. The White Rim Sandstone is of eolian origin, marking a near shore dune. Photograph taken from the Shafer Trail in Canyonlands National Park (Island in the Sky District).

by the Toroweap Formation, exposed below the rim of the Grand Canyon. Its three major members – a lower sandstone and evaporite interval, a middle limestone unit, and an upper evaporite and red bed interval – reflect a range of shallow marine, tidal flat, and sabkha depositional environments (Turner, 1990). Along the eastern shoreline an extensive dune field developed, comprising the lower part of the White Rim Sandstone.

Yet again, and for the final time during the Paleozoic, a marine transgression flooded part of the continental margins. Along the western margin, deposits of the transgression form the present Kaibab Formation, consisting of limestone, dolostone, sandstone, mudstone, and even evaporite deposits. Approximately equivalent rocks along the southern continental margin make up the San Andres Formation, dominantly a limestone. The Kaibab lithologies record a complex and temporally changing shallow marine environment (Hopkins, 1990). Along the Grand Canyon, the Kaibab Formation conformably overlies the Toroweap, forming the caprock of the southern rim and adjacent areas. Most of the prominent White Rim Sandstone (Fig. 5.16), which crops out over wide areas of Canyonlands National Park (Utah), originated as a coastal dune field along the eastern margin of the Kaibab sea (Condon, 1997). Eventually the sea spread over the White Rim dune field, reworking the top of the unit, burying and preserving it (Huntoon and Chan, 1987; Chan, 1989).

5.6 El Capitan

Slowly, the seas withdrew from Pangea, marking the end of the Absaroka transgression. But the decline in sea level was gradual, even halting. In the south and west, the seas lingered, continuing to deposit the Kaibab Formation. In a complex structural basin – the Permian basin – on the continental platform of southeastern New Mexico, West Texas, and northern Mexico, an inland sea formed, connected to the Permian ocean through a narrow channel (Fig. 5.17). The Permian basin was subdivided by the Central Basin Platform. In the western part of the Permian basin, known as the Delaware basin, conditions were favorable for development of an extensive and long-lived reef. Nearly 650 km long, the Capitan reef almost completely ringed the basin (Fig. 5.17). The fossil reef is magnificently exposed in the Guadalupe Mountains of Texas and New Mexico. The Capitan

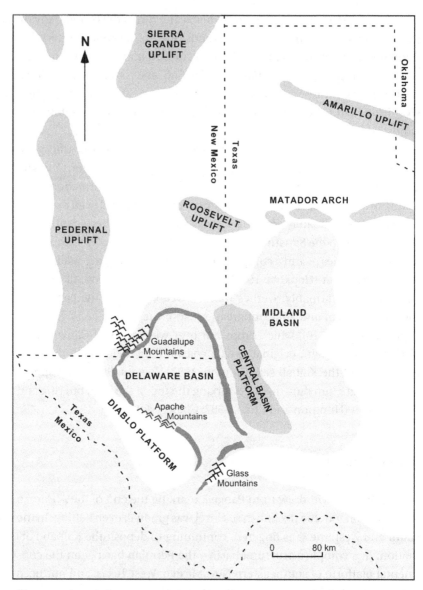

Fig. 5.17. The paleogeography of the Permian basin and surrounding structural features. The Capitan reef, shown by the dark gray pattern, surrounded the Delaware basin. Mostly buried, the reef is exposed in the Guadalupe, Apache, and Glass Mountains. From Jagnow and Jagnow (1992).

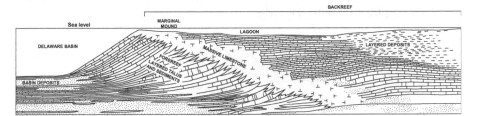

Fig. 5.18. An idealized cross section of the Capitan reef as it appeared in the Permian, with its depositional zones and present stratigraphy indicated. Modified from Jagnow and Jagnow (1992).

reef separates carbonate rocks of the shelf margin from deep-water basin deposits.

As with modern reefs, the Capitan reef was composed of colonies of calcite-secreting organisms that lived along the margin of the shallow plat-form. The skeletal remains of the organisms provided the framework for the reef (Fig. 5.18), which was filled in and cemented by skeletal debris and by crystals of calcium carbonate precipitated from the sea water. In con-trast with modern reefs, which are composed mainly of coral, reef-building organisms of Permian time were mainly calcareous algae, bryozoans, and sponges (Fig. 5.19) (Fagerstrom and Weidlich, 1999). It is likely that these organisms preferred slightly deeper and therefore relatively quieter water than those of modern reefs. The Capitan reef may have been as shallow as 9–15 m below sea level at certain stages of its development (Bebout and Kerans, 1993). The reef built a 'marginal mound' (Fig. 5.18). A talus slope, composed of broken pieces of reef, lime mud, carbonate sediment, and the fossils of other organisms, formed on the seaward side of the reef: the 'forereef.' On the landward side of the reef, a shallow lagoon developed between the marginal mound and the shore. In the warm, tropical waters of southern Pangea, gypsum was precipitated. Toward the shore, the gypsum interfingered with mud and sand laid down on broad tidal flats developed along the low coast line; toward open water, the evaporite deposits were interbedded with carbonate of the reef. These sediments now comprise the Yates and Tansill Formations (Fig. 5.20).

In response to rising sea level, the reef grew upward and outward. When it could not build upward, it grew outward over the talus and debris of the forereef (Fig. 5.21). At times of falling sea level, the reef became exposed, forming a series of islands. Overall, the reef grew for perhaps five million years, forming an edifice over 700 m thick.

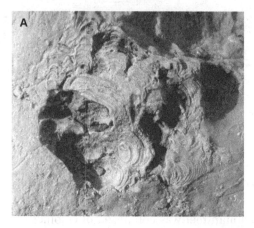

Fig. 5.19. Flora and fauna of the Capitan reef. (A) *Collenella* (an alga) and small sponges. *Collenella* head (laminated structure) is approximately 9 cm in diameter. (B) An upright sponge and a single fusulinid, *Polydiexodina*. Length of sponge approximately 18 cm. (C) Fusulinid-rich beds of the forereef talus. The fusulinid is identified as *Polydiexodina*. Photos A–C are all from the Permian Reef Geology Trail, Guadalupe Mountains National Park, Texas. Locations and identifications from Bebout and Kerans (1993). (D) Bits and fragments of delicate crinoid stems from the reef complex, 14 km southwest of Carlsbad, New Mexico (Adams *et al.*, 1993). Coin is about 2.1 cm in diameter.

In the deep waters of the Delaware basin, thick turbidite deposits were laid down on broad submarine aprons that built basinward. The resulting layers of sandstone and siltstone are characterized by graded bedding, ripup clasts, cross bedding, ripple marks, and soft-sediment deformation. Numerous ephemeral channels were cut into the surfaces of the fans as

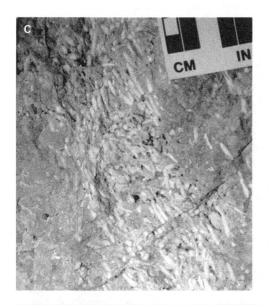

Fig. 5.19. (*continued*)

debris flows churned down the slopes. The deposits contain algae, bryozoan, fusilinids, and coral, forming thin layers of black, organic-rich limestone in the anoxic conditions of the basin.

Eventually, the connection of the Delaware Sea to the Permian ocean became more and more restricted. By the Late Permian, evaporitic conditions prevailed, and the basin filled with up to 600 m of anhydrite, limestone,

Fig. 5.20. (A) Back-reef limestone of the Yates and Tansill Formations of southern New Mexico. Horizontal bed below middle of photo is siltstone at top of the Yates Formation, overlying dolostone. Stromatolite structures (distorted bedding separated by vugs) are visible in Yates dolostone at lower left and the lower right of the photograph. (B) Detailed photo of fenestrally laminated (vugs elongate parallel to stratification) columnar stromatolites, indicative of relatively shallow water behind the reef (Bebout and Kerans, 1993). Location is 16 km southwest of Carlsbad, New Mexico (Adams *et al.*, 1993).

Fig. 5.21. The magnificent El Capitan in the southern Guadalupe Mountains, Texas, viewed from the east, exposes forereef talus beds dipping toward the south (left). Siltstone and sandstone of the Permian basin (Brushy Canyon Formation) underlie the slopes beneath the reef limestone. See also Plate 11.

and halite: the Castile Formation. The filling was slow, perhaps seasonal, forming laminated couplets of anhydrite and dark, organic-rich calcite (Fig. 5.22) interbedded with massive to poorly laminated halite. Carbonate and sulfate minerals were precipitated from evaporating surface water and subsequently rained down over the basin floor as a thin cover. To preserve the delicate laminae, the depth of sedimentation must have been generally below wave base, but at times water depth was shallow enough to allow crusts of gypsum to grow directly on the basin floor.

As was the case with the Paradox Formation, the overall compositions of evaporite minerals, especially the low proportion of halite relative to anhydrite, could not have been deposited by simple evaporation of sea water. While a connection to the sea, with direct marine inflow as a source for the brine, during Castile time is possible, several factors including seasonal events suggest a nonmarine basinal setting (Anderson, 1993). One possible model is that, during deposition of the Castile Formation, the basin was hydrologically closed, completely cut off from the sea and subject to fluctuating water levels. During periods of low water levels, chlorine-rich ground water, possibly derived from sea water and/or from previously

A

B

Fig. 5.22. For caption see opposite.

deposited evaporites, seeped into the basin from the south. During periods of high water, recharge by meteoric water, entering the basin from the west, dominated. These (presumably seasonal) 'freshening' recharges resulted in rapid, synchronous, and basin-wide accumulation of calcite laminae.

Finally, the basin was completely filled and the extensive reef complex buried by evaporitic minerals, siltstone, sandstone, and mudstone (Salado and Rustler Formations). Under burial conditions, the gypsum dehydrated to anhydrite.

5.7 Sonoma orogeny

At the latitude of southern California, all of the tectonostratigraphic elements of the Cordilleran margin are abruptly truncated to the southwest. Suspected pieces of the margin now lie in northern Baja California and along the west coast of mainland Mexico, apparently offset from miogeoclinal and platform rocks in Nevada and eastern California by left-lateral displacement. Thus, it is likely that transform tectonics may have affected parts of this margin during the Pennsylvanian (Stevens *et al.*, 1992). Also in the Pennsylvanian and Permian, concurrently with deformation of the Ancestral Rocky Mountains, an intraoceanic magmatic arc, including the

---

Fig. 5.22. (A) Couplets of white to gray anhydrite laminae (1.5–6 mm in thickness) alternating with sub-millimeter-sized, dark brown, organic-rich calcite laminae characterize much of the Castile Formation. Castile sediments were deposited in the Permian basin after its connection to the sea became severely restricted. Couplets possibly represent seasonal varves. Despite numerous studies, the origin of the small-scale folds that characterize the Castile and render its appearance so intriguing remains problematic. Probably they result from post-depositional tectonic deformation, but the various factors that control folding have not been determined (Watkinson and Alexander, 1993). Location is 46 km southwest of Carlsbad, New Mexico (Austin *et al.*, 1993). (B) Beautiful, large crystals of selenite in the Castile Formation, some longer than 5 m, were formed as ground water locally hydrated anhydrite to gypsum. The precise controls on ground water flow and the mechanism of hydration are not well understood. Pen is 14.5 cm long. Locality is 43 km southwest of Carlsbad (Austin *et al.*, 1993).

microcontinent Sonomia, lay offshore of the continental margin along most of the present USA. Separated from the exposed continent by a sedimentary basin (the Havallah basin), it approached from the west as North American lithosphere descended below it (Dickinson, 1981). In Southern California the arc crossed obliquely onshore to form a margin similar to that of the present Andes Mountains of Central and South America. One possible explanation for the truncation and apparent left-lateral offset of the margin is that the left-slip zone was a transform boundary joined to the westward-dipping subduction zone. Alternatively, the left-slip boundary was a left-slip fault zone lying inboard of an obliquely convergent west-dipping subduction zone that lay along the entire western margin of the Cordillera (Burchfiel *et al.*, 1992).

In the Late Permian, orogeny and volcanism along the continental margin were renewed in another spasm of contractional deformation that closed the Havallah basin. Deformation, described in more detail in the next chapter, began in the Late Permian and continued into the Triassic, not noticing the mass extinction event (Box 5.1) at the boundary. Late Permian time also witnessed the withdrawal again of the seas from most of western North America, marking the end of deposition of the Absaroka sequence. Although, in the Cretaceous, part of western North America, including the Colorado Plateau, was part of a great interior seaway that stretched eastward into the present Great Plains, the seaway was connected to the global ocean to the north and south but not to the west (Chapter 7).

Box 5.1 End-of-Permian extinction The Capitan reef was unlike modern reefs in that calcareous algae, bryozoans, and sponges were the main reef-building organisms. In contrast, modern reefs are built mainly by corals. The change in reef building was only one result of a major extinction event – a severe global ecological crisis that affected plants and animals on land and in the oceans – marking the end of the Paleozoic Era approximately 251.4 Ma (Bowring *et al.*, 1998). Although numerous mass extinctions are recognized throughout the geologic record, the event at the end of the Permian Period is unquestionably the greatest of all known extinctions. Life on Earth came very close to being completely extinguished. Up to 96% of all marine species were wiped out. Thirty two families of tetrapods disappeared (Colbert, 1995), and altogether about 70% of terrestrial vertebrate families became extinct (Erwin, 1994). Peat-forming plants also became extinct, resulting in a 10 Myr-long gap in formation of coal. Full recovery to pre-Triassic levels of coal formation required 50 Myr (Retallack *et al.*, 1996). Reef building also ceased completely for some 12 Myr. Extinction caused an abrupt shift in carbon isotope composition (Musashi *et al.*, 2001), which lasted for 50–100 ka. Reduced diversity of marine organisms resulted in reduced demand for carbon for photosynthesis, resulting in buildup of dissolved CO_2 in marine water and increased partial pressure of CO_2 in the atmosphere in the earliest Mesozoic. Yet, possible global warming resulting from increased atmospheric CO_2 has not been substantiated for this increase (Wang *et al.*, 1994). The abrupt decline in species at the boundary is also correlated with a sharp increase in fungal remains (Eshet *et al.*, 1995).

In southern China, the boundary is associated with a concentration of Fe–Si–Ni grains and kaolinite, and with a sharp decrease in $^{34}S/^{32}S$ and $^{87}Sr/^{86}Sr$ (Kaiho *et al.*, 2001). In the Karoo basin of South Africa, the Permian–Triassic boundary marks an abrupt change in sedimentation from meandering to braided river systems, from which a major and probably global die-off of rooted plant life is inferred (Ward *et al.*, 2000). However, the global significance of the data in these two studies has been sharply criticized (Koeberl *et al.*, 2002; Rees, 2002). Although they may constitute evidence for a major world-wide event at the Permian–Triassic boundary, similar studies at other exposures and a better understanding of the compositional processes involved need to be undertaken before these ideas are generally accepted.

The cause of this greatest extinction remains elusive. Many hypotheses have been advanced, including eruption of 2×10^6 km^3 of basaltic lava flows in Siberia, the impact of large bolides, abrupt climate change, and major changes in sea level. A contributing factor may have been the assembly of Pangea, which led to an average thickening of continental crust. The elimination of many shallow-marine environments, in which marine flora and fauna flourished, may have significantly stressed marine organisms. No single explanation seems compelling at present, but circumstantial evidence, such as that cited above, is compatible with an impact origin. Although the cause or causes of the extinction remain enigmatic, evidence provided by recent age-dating studies in China and by sedimentological studies in South Africa indicate that it was sudden, occurring over a period of 50 ka or less (Smith and Ward, 2001).

The overall result of the Sonoma orogeny was accretion of new terranes to the western margin of the continent. Recognition of the addition of 'exotic' or 'accreted' terranes to the seaward edge of the continent, and an understanding of the process by which accretion occurs, were achieved largely in western North America. Appreciation of the role and importance of accretion to crustal growth represents one of the major intellectual achievements of geology in the twentieth century. The concept of accretion of island arcs and related terranes, first worked out for Phanerozoic rocks, is now applied to Precambrian rocks to understand formation of continental crust (Chapters 1–3).

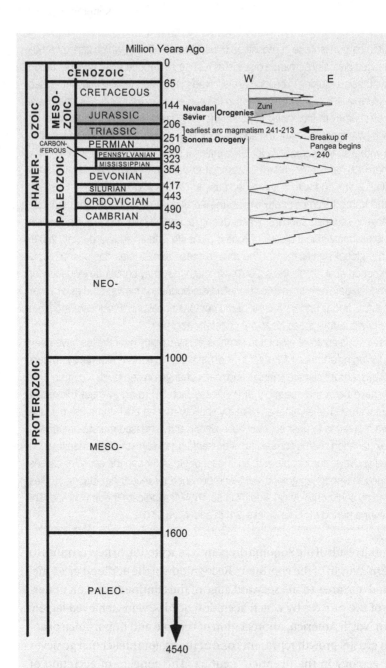

Geologic time scale. For explanation see p. 3.

Deserts and dinosaurs

Triassic and Jurassic

The massive Jurassic Navajo Sandstone (light-colored, cliff-forming unit) overlies the Kayenta Formation along the Burr Trail over Waterpocket fold at Capital Reef National Park, Utah. Together with the laterally equivalent Nugget and Aztec Sandstones, these formations comprise one of the largest ancient sand dune deposits preserved on Earth, stretching over an area of more than 150,000 km².

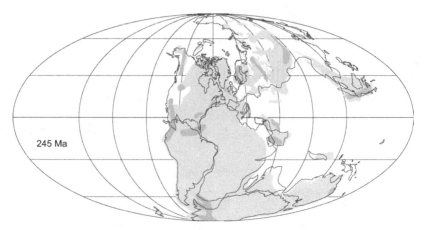

Fig. 6.1. Paleogeographic map of Pangea for the Early Triassic, approximately 245 Ma. Stippled pattern indicates land areas, with darker pattern indicating higher elevation. Modified from Smith *et al.* (1994).

6.1 Setting

By the beginning of the Mesozoic, Pangea was nearly fully assembled. Only a few microcontinents, such as North and South China, remained yet to collide with the main Pangean landmass, which they did by the end of the Triassic (Rees, 2002). Stretching nearly from pole to pole, Pangea drifted slowly northward across the equator (Fig. 6.1). The western margin of the supercontinent and of present North America continued to be a collisional boundary during the Mesozoic era, with the Cordilleran miogeocline present along the continental margin. Great tracts of new continental crust – 'exotic terranes' – were accreted to the continent during the Mesozoic. Together with magmatic arcs and sediments of the adjoining trenches, these terranes formed what are now the States of California, Oregon, and Washington, and much of Nevada and Arizona. The Early Triassic also saw the beginning of the end of Pangea, as great continental blocks were carved from Pangea to form new plates. The demise of Pangea was not sudden, but rather was drawn out over the next 150 Myr, not to be completed until the early Tertiary (*c.* 45 Ma). However, the disassembly of Pangea did not affect the western part of Pangea, including the present Southwest.

The edge of the craton extended through central Utah and southern Nevada. To the west, in central Nevada, lay terrane accreted to the continent during the Antler orogeny, and the miogeocline. On the continental margin, shallow-marine sediments alternated with terrigenous deposits as the shoreline fluctuated. Farther to the west, in western Nevada and California, deep-water sediments were deposited on oceanic crust throughout the Mesozoic. From the Early Triassic to the Early Jurassic, the shoreline withdrew from central Utah to western Nevada. Then, beginning in the middle Jurassic, marine waters again flooded onto the craton in the Zuni transgression. Later transgressions occurred at frequent intervals thereafter, partly in response to deformation along the western margin of Pangea.

As discussed in the previous chapter, most of the orogenic activity of western Pangea was confined to the margin of the supercontinent. Only relatively minor mountain-building affected the continental margin farther inland. In contrast with the major orogenic events occurring in other parts of Pangea, most of the Southwest was tectonically quiescent, apparently on the sidelines of most of the activity. Continental conditions prevailed widely throughout the Southwest during much of the Mesozoic. On the Colorado Plateau, Mesozoic strata (largely Triassic and Jurassic) comprise the best exposed and best documented desert sediments in Earth's stratigraphic record (Allen *et al.*, 2000). Sediments were largely derived from the Ancestral Rocky Mountains to the northeast or from new highlands to the south. Parts of the Southwest during the Mesozoic were covered by shallow seas, as deformation-induced subsidence and relative rise in sea level allowed marine waters to spread eastward onto the continent. Strata deposited during this period of time are widespread over much of the Colorado Plateau, and are responsible for the multihued landscapes and spectacular scenery displayed in the many national and state parks and monuments of the region. Because continental conditions prevailed widely throughout the Southwest, the great extinction event marking the Paleozoic–Mesozoic boundary (see Box 5.1) is almost not preserved in the Southwest. Yet, the depositional environment of the Southwest, whether continental or shallow-marine, was entirely controlled throughout this interval by the convergent-margin tectonic events of the western plate boundary.

Although convergent tectonics dominated along the western margin of Laurentia from the Late Devonian Period (Chapter 4), the oldest evidence for arc magmatism in the Southwest is provided by a series of Early Triassic plutons (241–213 Myr old) stretching from central California and western

Nevada to northern Sonora (Barth *et al.*, 1997). The arc may have comprised the highland region stretching across southern Arizona, which is inferred from a south-derived source for volcanic clasts in the Chinle Formation (Stewart *et al.*, 1986). These volcanoes were probably also the source for ash, which is characteristic of parts of the Chinle Formation (discussed below).

6.2 Orogeny and magmatism: the early arc

Despite the fact that stable conditions *generally* prevailed throughout the Southwest, the region did not completely escape mountain-building activity. Geologists recognize two orogenies during the Triassic and Jurassic in the span of approximately 100 Myr. Although recognized and named separately, viewed in a geologic time frame these individual orogenies are just pulses of a single, nearly continuous mountain-building event referred to as the Cordilleran orogeny. But important at they were, these orogenies were but a prelude for the big one to follow, the Laramide orogeny, which penetrated deeply into, and profoundly affected, the craton (Chapter 7).

Early Mesozoic orogeny was actually a continuation of the Sonoma orogeny, which, as discussed in the previous chapter, began before the close of the Permian. As was the case for the Antler orogeny, the Sonoma orogeny occurred as the continental margin was drawn into the west-dipping Sonoma subduction zone (Fig. 6.2). The intervening basin was

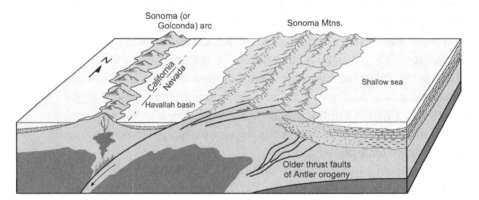

Fig. 6.2. Cross section through the Sonoma (or Golconda) arc and Golconda thrust. Approximate location is shown in Fig. 6.3. Modified from Levin (1991).

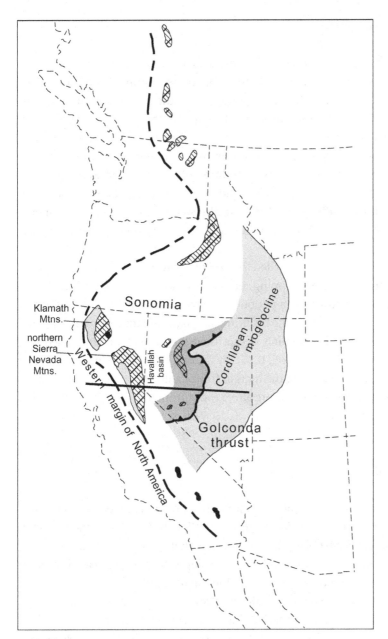

Fig. 6.3. Tectonic elements of the Sonoma orogeny. Heavy line is location of cross section shown in Fig. 6.2. Dark shading is offshelf and marginal basin sediments of the Havallah basin; medium gray is late Paleozoic miogeoclinal sediments, and sediments overlapping the Antler orogen. Black pattern indicates Permian plutons; ruled pattern is volcanic rocks; and light gray indicates sediments of the western melange. Modified from Burchfiel *et al.* (1992) and Blakey (1989).

closed and the sediments pushed onto the continental margin. This process added the Sonoma arc (Sonomia) to the continent, and formed a thrust-faulted and uplifted highland region, the Sonoma Mountains (or Golconda allochthon), stretching through central Nevada (Fig. 6.3). In Nevada, oceanic rocks were pushed 50 km eastward along the Golconda thrust zone onto the eroded structures of the Antler orogen, forming a series of thin fault slices (Fig. 6.2). Addition of Sonomia and welding of the accretionary wedge of sediments added some 300 km of new crust to the continent (Burchfiel *et al.*, 1992). To the east of the Sonoma Mountains lay a shallow basin. Sedimentation in the basin buried the earlier Antler orogen.

The Sonoma orogeny spelled the end of westward-dipping subduction. Shortly thereafter, east-dipping subduction was initiated, possibly following or concurrently with a period of transform faulting, which truncated the continental margin. The resulting magmatic arc that formed along the continental margin obliquely overprinted the Proterozoic and Paleozoic sedimentary belts (Ingersoll, 1997). By the end of the Triassic, a line of plutons was emplaced along the margin from California to northern Mexico, marking the position of the arc along the continental margin south of the Sonoma orogenic zone (Barth *et al.*, 1997).

6.3 More red beds

Along the western edge of the continent, deposition of marine strata continued into the Early Triassic. In the shallow basin east of the Sonoma Mountains (eastern Nevada and western Utah), a typical shelf sequence of limestone, shale, and sandstone was deposited, grading eastward into sandier marine sediments and continental red beds.

In contrast, subaerial conditions prevailed throughout most of the Southwest during the Triassic. The continental conditions that had become firmly established during the Permian continued into the Mesozoic. The oldest of the Triassic red beds, the Early Triassic Moenkopi Formation (Fig. 6.4), consists of red and brown mudstones deposited largely on tidal mudflats adjacent to the western ocean. Mudflat deposits are overlain by a sequence of fluvial and lacustrine sediments, comprising the Upper Triassic Chinle Formation and equivalent strata. The Chinle Formation

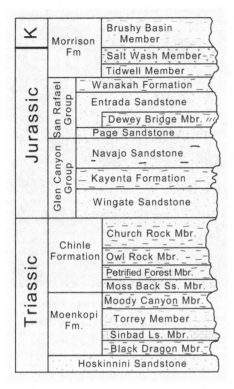

Fig. 6.4. Stratigraphic column of Mesozoic rocks from the Canyonlands area of southeastern Utah. From Hintze (1988).

consists largely of sandstone, mudstone, and minor freshwater limestone deposited in fluvial channels and on adjacent floodplains and mudflats; in deltaic distributary channel systems; and in lakes, marshes, and deltas on a broad lowland – an immense alluvial plain – developed on the margin of the continent. Generally, the Chinle Formation fills large paleovalleys incised into the underlying Moenkopi Formation. The basal unit of the Chinle Formation, the Shinarump Member (Fig. 6.5), and equivalent rocks to the north and east provide evidence for a major, northwest-directed river system (Fig. 6.6) stretching from a source in the Amarillo–Wichita uplift of northern Texas to the Cordilleran continental margin in central Nevada (Riggs *et al.*, 1996). The river traversed a topographically low continental interior, probably punctuated with marshes and lakes, before debouching into the western ocean. Locally, relict uplifts remaining from the Pennsylvanian Ancestral Rockies orogeny – the Uncompahgre, Front Range, Pedernal and other uplifts – and the growing Mogollon uplift to the south

Fig. 6.5. Cross-bedded conglomerate of the Shinarump Member of the Chinle Formation, near Holbrook, Arizona. The Shinarump, which represents paleovalley fills, accumulated as diffuse gravel sheets and on longitudinal bars along braided streams. Paleocurrent indicators consistently indicate a source to the south (the Mogollon highlands), possibly as close as 190 km. Rock types present as clasts include Precambrian quartzite and volcanic rocks, and Paleozoic sandstone and limestone (Blakey and Middleton, 1986). Scale is 51 cm.

contributed clastic debris to the drainage system. The large river system probably persisted until late Chinle time. In general, the climate during deposition of the Chinle is interpreted as tropical and characterized by humid conditions, owing presumably to the low latitudinal position of the Southwest at that time (Fig. 6.1).

Much of the Chinle Formation contains abundant volcanic detritus, ranging in size from ash to cobbles derived from southerly sources (Blakey and Middleton, 1986; Riggs *et al.*, 1996). In particular, the delicately colored silicified logs that characterize the Petrified Forest Member of the Chinle owe their spectacular preservation to silica leached from volcanic ash (Fig. 6.7). Logs are dominantly of the conifer *Araucarioxylon arizonicum*, related to the modern Norfolk Island pine. The logs were rafted to their present locations from conifer forests in the volcanic highlands to the south in massive floods and, stripped of their limbs and leaves, stranded

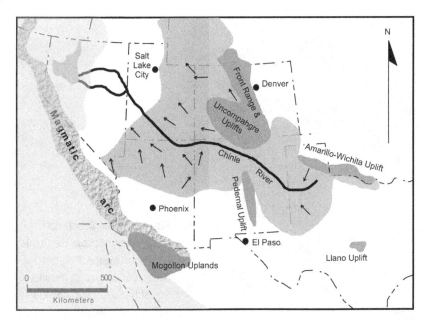

Fig. 6.6. A transcontinental river system, perhaps similar to the modern Mississippi River, traversed the Southwest in Middle Triassic time (224–221 Ma). The river may have had its source in the Amarillo–Wichita uplift area of northern Texas, debouching in the paleo-Pacific Ocean in present-day western Nevada. From Riggs *et al.* (1996).

on sandbars or trapped in logjams on flood plains. Some logs exceed 60 m in length. Volcanic ash from the magmatic arc to the south and southwest, incorporated into bounding sediments, provided an abundant source of silica (SiO_2) for silicification of the wood. Evidence from the Chinle Formation indicates the presence of a significant volcanic highland area to the south and west, part of the early Mesozoic magmatic arc system. The depositional environment of the Petrified Forest Member is interpreted to indicate that the swamps periodically dried up. Precipitation was very likely seasonal, with alternating wet and dry periods.

As the Absaroka seas slowly withdrew from the western margin, deposition of continental red beds comprising the Chinle Formation spread westward across eastern Nevada and western Utah, the area of the former shallow marine basin. In Utah, the remnants of the formerly imposing Uncompahgre uplift were nearly buried by sediments of the Chinle Formation.

Fig. 6.7. Volcanic ash-rich sedimentary rocks of the Chinle Formation overlie a log-bearing horizon at Long Logs, Petrified Forest National Park, Arizona. Some of the silica (SiO_2) responsible for preservation of the wood may have been derived from the ash in these strata.

6.4 Early dinosaurs

The Chinle Formation provides evidence for one of the earliest dinosaurs on Earth, the theropod *Coelophysis buri*. Remains of a few individuals have been found in the Petrified Forest Member of the Chinle Formation at Petrified Forest National Park in Arizona. However, at Ghost Ranch in northern New Mexico, the remains of at least a thousand individuals have been collected (Fig. 6.8), and many thousand more may remain. At this site, the fossils are found in the informally named 'siltstone member,' stratigraphically equivalent to the Petrified Forest Member in Arizona. The small carnivore *Coelophysis*, ranging in length to over 3 m, is the oldest North American dinosaur for which whole skeletons are recovered. How did this great number of individuals come to be concentrated at the Ghost Ranch site? The main bone-bearing strata of the siltstone member comprise abandoned channel sediments deposited on the floodplain of a stream or river.

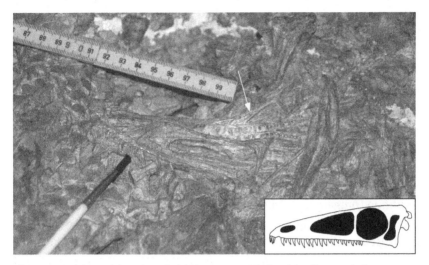

Fig. 6.8. Amidst various other skeletal parts, two skulls of the theropod dinosaur *Coelophysis* are visible in this block of Triassic Chinle Formation (siltstone member). For comparison, inset shows restoration of a *Coelophysis* skull (Colbert, 1995). Bristles of brush point to one of several teeth in the upper jaw of the lower and larger skull. Second skull is indicated by arrow. *Coelophysis*, up to 3 m in length or slightly longer, is the oldest North American dinosaur for which whole skeletons are recovered. This eight-ton block was removed from a quarry at Ghost Ranch, New Mexico, in 1985 and now resides in the Ruth Hall Museum of Paleontology at Ghost Ranch. Probably thousands of skeletons are present at this site, swept into a 'logjam' by raging floodwaters. Scale is in cm. Photograph by permission.

The abundance of skeletons, their positions, and the overall depositional setting, are interpreted to suggest that the dinosaurs died elsewhere and were rafted to the quarry locality by floodwaters, clogging and filling a small distributary channel. But what actually killed these thousands of animals? Although difficult to say, it seems most likely that a *Coelophysis* herd perished during a period of extended drought, as has been observed for modern reptiles, and were then swept together by flooding. This environmental model of drought and floods agrees with evidence indicating that the climate during early Chinle time appears to have been monsoonal, with seasonally wet and dry periods (Schwartz and Gillette, 1994; Colbert, 1995).

By the end of Chinle deposition, the climate was becoming increasingly arid, paving the way for the great deserts of the Jurassic Period.

6.5 Seas of sand

For Pangea, the Jurassic was a time of great tumult. Africa tore away from North America, rending the supercontinent apart to form the widening Atlantic Ocean. Along the western margin, collisions with incoming island arcs and other tectonic flotsam accelerated. Collisional processes, becoming increasingly important near the end of the Triassic or early in the Jurassic, deformed and uplifted the pre-Jurassic strata. As a result, the Chinle Formation pinches out toward the west, and Lower Jurassic strata overstep progressively older strata toward the arc terrane (Fig. 6.9) (Marzolf, 1988). Continued orogeny throughout the Jurassic (discussed later in this chapter) and the load that it superimposed on the continental margin, depressed the crust downward, creating a topographic depression in which the thick

Fig. 6.9. Close to the volcanic arc, a substantial angular unconformity records late Triassic to Early Jurassic deformation along the continental margin. Seen in this photograph of an outcrop in the Cowhole Mountains, in the Mojave Desert of California, the base of the Jurassic section is marked by a limestone breccia containing angular blocks of the underlying Pennsylvanian–Permian shelf limestone (gray) in a matrix of reddish mud. The breccia fills a broad channel cut into the underlying carbonate terrane. Above the breccia is the Moenave–Kayenta Formation, a terrestrial fluvial deposit (Marzolf, 1988). Scale is 41 cm.

eolian and other continental deposits of the Jurassic Period accumulated (Allen *et al.*, 2000). Marine waters were able to invade, and for parts of the Jurassic, much of the central and northern Rocky Mountain region was covered by a giant saltwater bay. The Jurassic section is punctuated by seven major (i.e. regional) and several lesser unconformities, each probably representing a pulse in orogenic activity. The unconformities are important in helping to define and correlate 'packages' of continental strata.

Increasing aridity, begun during Chinle times, meant that for much of the Southwest, the Jurassic was also a time of great deserts. By Early Jurassic time, the area of the present western USA lay in the tradewind belt north of the equator: probably 13°–16°N by the Middle Jurassic (Armstrong, 1995). Blowing offshore persistently from the northeast, the winds kept the western edge of the continent as dry as the modern Sahara Desert. Enormous seas of sand tens of thousands of square kilometers in extent occupied coastal regions adjacent to a shallow sea (Fig. 6.10; see also Plate 3). Ergs

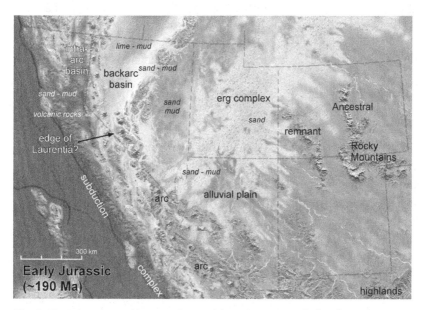

Fig. 6.10. Paleogeographic map of part of the western USA during the Early Jurassic Period (approximately 190 Ma). The Southwest lay in the tradewind belt. Winds blowing persistently from the northeast kept the southwestern margin of the continent as dry as the modern Sahara Desert, and sand seas occupied much of the region. Modified from R. C. Blakey, website: http://vishnu.glg.nau.edu/rcb/paleogeogwus.htm, by permission. See also Plate 12.

were intermittently replaced by fluvial and minor lacustrine environments, leaving deposits of siltstone and stream-laid sandstone interstratified with erg deposits. What happened to create these great dune fields? Was there a change in the global climate, or was the spread of the sands controlled by local factors? Whatever it was, the change in deposition probably had to do with more than Pangea's slow and gradual drift northward.

In the first of these enormous **sand seas** (Box 6.1), the widespread Lower Jurassic Wingate Sandstone accumulated (Fig. 6.11), gradually spreading

---

Box 6.1 Sand sea and sabkha The term sand sea refers to extensive regions of desert covered with shifting sand and characterized by complex dunes. Because the sands of the Sahara Desert are a good modern analogue for many sandstone deposits in the geologic record, the synonymous term **erg**, derived from the Arabic of northern Africa, is also widely used. Some 1,360,000 km² of the Sahara Desert are covered by sand seas (Harris and Levey, 1975).

Derived from the Arabic word 'sebkha,' for 'salt flat,' the term **sabkha** is used to describe a broad supratidal or intertidal flat developed along the margin of an arid landmass. Sabkhas are typically saline, and may be occupied by marshes or shallow lakes after rain. Evaporite minerals may include anhydrite, gypsum, and dolomite. The salt flats developed along the Persian Gulf, whence the term arises, are examples of sabkhas (Jackson, 1997; Prothero and Schwab, 1996).

All of these terms imply an arid environment, characterized by infrequent and probably seasonal precipitation.

---

over the complex drainage system of the Chinle Formation. Winds scoured the mudflats and floodplains of the Chinle landscape, whipping the sand and silt grains into large dune fields. Then, for a brief time, mainly in the area of the western Colorado Plateau, deposition of dune sands was interrupted, as stream deposits buried the Wingate dunes. These deposits, the Moenave and Kayenta Formations (Fig. 6.12), are mainly sandstone and mudstone derived from Precambrian sources in Colorado (the relict Uncompahgre highlands) and volcanic sources to the south. The Kayenta has yielded the bones of dinosaurs and small mammals that lived along the water courses.

Again, the dune sands returned. The second of the great sand seas is represented by coastal and inland dune fields of the Navajo Sandstone (Chapter 6 frontispiece) and the equivalent Aztec Sandstone of Nevada and southeastern California (Fig. 6.13), and possibly of the Nugget Sandstone of Utah and Wyoming (Hintz, 1988). Arid and windy conditions

Fig. 6.11. The massive, cliff-forming Wingate Sandstone represents the earliest of the Jurassic sand sea deposits. The cliffs photographed here overlook the Shafer Trail in Canyonlands National Park, Utah.

prevailed, and the Navajo desert must have looked similar to the sand seas of the modern Sahara Desert. Together, these formations comprise one of the largest fossil sand-dune deposits preserved on Earth, stretching over an area of more than 150,000 km^2, mainly in Utah and northern Arizona (Fig. 6.14). Generally, the erg lay close to the continental margin, hence the Navajo Sandstone is not present in New Mexico and southern Colorado (Fig. 6.15). Most of the Navajo Sandstone consists of a white, nearly pure quartz sand. The deposit ranges to more than 600 m thick in Zion National Park. It displays large-scale, high-angle cross stratification in conspicuous wedge-shaped sets typical of dune deposits. Individual foreset beds range up to 19 m in height. Dip directions of crossbeds indicate wind direction from the northwest.

Major dunes or dune complexes are separated by nearly horizontal first-order **bounding surfaces** (Box 6.2), interpreted to have formed by the

Fig. 6.12. At Valley of Fire State Park, Nevada, the contact between fluvial sandstone, siltstone, and mudstone of the Moenave–Kayenta Formation (undifferentiated) and eolian dune deposits of the overlying Aztec Sandstone is transitional. Seen in this photograph near the Park Visitor Center, cross-bedded sandstone layers (white) are interbedded in the upper Moenave–Kayenta Formation (Marzolf, 1988). See also Plate 13.

Box 6.2 Bounding surfaces Modern well-developed sand seas are characterized by rhythmically spaced dunes and interdune areas. The migration of interdune areas downwind creates erosion surfaces that truncate the cross-bedded dune deposits below. Horizontal strata immediately overlying the surfaces are interpreted to represent interdune deposits, and may include waterlaid deposits of ephemeral, small lakes. The most extensive of these surfaces are termed 'first-order' bounding surfaces (Fig. 6.16). A hierarchy of bounding surfaces is recognized based on their lateral extent and geometrical relationships (Kocurek, 1981).

Fig. 6.13. Cross-bedded fossil eolian sand dunes of the Aztec Sandstone at Valley of Fire State Park, Nevada. The Aztec Sandstone of Nevada and Utah is the equivalent of the Navajo Sandstone in northern Arizona and southern Utah. Together, the sandstone unit represents the largest erg recognized in the geological record.

migration of interdune areas across the dune field. Some first-order bounding surfaces are marked only by the erosional contact of dune foresets, but depressions on these surfaces may be filled with nearly horizontal, interdune deposits. Interdune deposits may include freshwater limestone and various sedimentary structures such as mud cracks and burrows. They may even contain vertebrate fossils and tracks. Many first-order

Fig. 6.14. Distribution of outcrops of the Navajo Sandstone (shaded area). From Winkler *et al.* (1991).

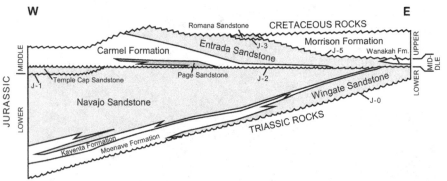

Fig. 6.15. Stratigraphic section of Jurassic rocks along the Arizona–Utah state line. This section shows the pinchout of the Navajo erg eastward from the continental margin (Fig. 6.14) and overstepping of the dune deposits by Middle and Upper Jurassic rocks. The J-0 through J-3 and J-5 unconformities are indicated. Modified from Peterson (1988).

bounding surfaces and associated interdune deposits can be traced laterally over distances of kilometers (Winkler *et al.*, 1991).

6.6 Orogeny: Nevadan and Sevier!

The Mesozoic was a time of frequent, if not continuous, orogeny along the western continental margin, beginning with the Sonoma, described

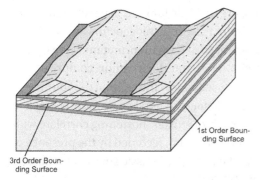

1st Order Boun-
ding Surface

3rd Order Boun-
ding Surface

Fig. 6.16. Model for migrating dunes and interdune areas (shaded) in an erg. Migration of interdune areas truncates dune deposits, creating first-order bounding surfaces. Second-order bounding surfaces (not shown), which are intermediate in scale, are formed by compound crescentic dunes. Third-order bounding surfaces are formed as sets of simple crescentic sand dunes migrate over, and thus truncate, prior sets. After Kocurek (1981).

previously. During the latest Triassic to Middle Jurassic time, a narrow magmatic arc became well developed along the entire continental margin, extending from Nevada, California, and Arizona as far south as northern Sonora and Chihuahua (Schweickert, 1976; Busby-Spera, 1988; Lawton and McMillan, 1999). The nearly continuous deformation of the margin was a result of strong, possibly oblique, plate convergence and of interplate coupling between overriding and overridden slabs (Burchfiel *et al.*, 1992). Initially, deformation of the continental margin was relatively minor, despite active plate convergence and intra-arc deformation. East of the arc, on the continental margin, was a broad area of shallow-marine and non-marine deposition, which was intermittently affected by the deformation associated with the convergence.

In the central Mojave Desert of southeastern California, eugeoclinal strata were thrust over a platform sequence along the truncated continental margin in the Middle Triassic to Early Jurassic, between 240 and 175 Ma (Miller *et al.*, 1995). In northern California and in the Great Basin area of Utah, crustal shortening resulting from accretion of multiple exotic terranes to the continental margin may have begun as the early as Early Jurassic, 200–185 Ma (Allen *et al.*, 2000). Major crustal shortening certainly occurred during the Middle to Late Jurassic, 163–143 Ma, and is referred to as the Nevadan orogeny. On the continental margin, a protracted period of compressional tectonism began in the Late Jurassic, in which areas more than 1000 km eastward through the Paleozoic miogeocline and well into the North American craton experienced deformation, metamorphism, and plutonism (Burchfiel *et al.*, 1992). This later deformation comprised the Sevier orogeny, which had major consequences throughout the Southwest. The Sevier, occurring mainly in the Cretaceous, is taken up more fully in the next chapter. Generally, these periods of crustal shortening correlate with pulses of arc-related volcanism (171–148 Ma), recorded by the presence of altered volcanic ash beds in Middle to Upper Jurassic strata of California, Utah, and Arizona (Kowallis *et al.*, 2001).

One possible explanation for at least part of the Nevada orogeny is collision of a westward-dipping, intraoceanic subduction zone with a remnant eastward-dipping, continental-margin subduction zone off northern California (Schweickert and Cowan, 1975; Ingersoll, 2000). In this model, a magmatic arc, underlain by an east-dipping subduction zone, developed on the continental margin during the Late Triassic. During the end of Middle Jurassic time, an east-facing island arc (underlain by a west-dipping subduction zone) developed some distance west of the continental margin. As the intervening lithosphere was consumed, the two arcs approached one another. Approximately 150 Ma, the two arcs collided, with subduction along the west-dipping zone ceasing. Crustal shortening of the continental margin resulted from partial subduction of buoyant forearc strata of the margin (Ingersoll, 2000).

The developing Sevier orogen, while largely outside of the Southwest proper, had a major effect on sedimentation in the Southwest during the Jurassic. Not only did it create a barrier to the ocean to the west, but also the weight of the thrust sheets depressed the crust, creating a shallow foreland basin to the east. This process is discussed more fully in the next chapter. Crustal loading was very important in that, by forcing the continental crust downward, it created a topographic depression in which the thick eolian and other continental deposits of the Jurassic Period could accumulate and be preserved (Allen *et al.*, 2000). In combination with the Zuni transgression, parts of the western USA were depressed below sea level. Slowly, the Jurassic sea invaded the basin. In doing so, it gradually inundated the vast sand sea that is the Navajo Sandstone, creating the narrow interior Sundance Sea (also called the Utah–Idaho trough) that stretched from the northern part of the continent to

New Mexico. From Utah northward, marine waters were connected to the open ocean. However, in southern Utah and New Mexico, an isolated basin only intermittently connected to the ocean developed. Around the margins of the basin, depositional environments ranged from tidal mud flats to eolian sand dunes. Sedimentary deposits grade into each other both laterally and vertically, resulting in complex local stratigraphic relationships and a plethora of formation names. Over wide areas of the Colorado Plateau, the Carmel Formation represents the first deposits of this shallow sea as it transgressed over the dunes. The Carmel Formation consists of shallow marine to continental red mudstone, sandstone, limestone, and gypsum, in part derived from source terranes to the south (Blakey *et al.*, 1996). In the western and northern parts of its area (e.g. northern Utah), the Carmel Formation is shallow-marine, grading into marine shale, mudstone, and fossiliferous limestone of the Arapien Shale and Twin Creek Limestone in the Great Basin (Peterson, 1988; Hintze, 1988; Blakey *et al.*, 1996). To the east, strata of the Carmel Formation were deposited in a continental environment. Here, they represent tidal mud flat and coastal sabkha environments adjacent to the sea. In northern Arizona and southern Utah, mudstone of the Carmel grades laterally into eolian sandstone of the Page Sandstone (Blakey *et al.*, 1996). The Page is a coastal and inland erg deposit consisting of cross-stratified dune sandstone with interbedded red, silty sabkha deposits. Large-scale erosional and depositional patterns in the Carmel Formation and Page Sandstone can be correlated with marine carbonate and siliciclastic deposits of the Twin Creek Limestone to define broad depositional events along the continental margin (Blakey *et al.*, 1996). In places, the Carmel Formation is highly contorted, perhaps owing to loading of the overlying sediments.

Along the western margin of the embayment, a large erg and sabkha complex developed over the present Colorado Plateau (Fig. 6.15). The shifting sands of the erg became the Entrada Formation, which is variable in its characteristics. The Entrada Formation is correlated, in part, with the Cow Springs Formation, which may simply represent a separate part of the overall Entrada dune field fed by sand from a different source. Mainly a massive, cross-bedded eolian sandstone (Figs. 6.17, 6.18), the Entrada consists of a central erg facies of large, crescent-shaped dunes (Fig. 6.19) and enclosed interdune areas, and a coastal dune facies of smaller, more irregular

Fig. 6.17. Part of the Jurassic section at Red Mesa, north of Mexican Water, Arizona. At this location the lower part of the Summerville Formation is characterized by contorted strata. Stratigraphic identification from S. Lucas and S. Sempken (personal communication). See also Plate 14.

Fig. 6.18. The Entrada Sandstone, near Moab, Utah.

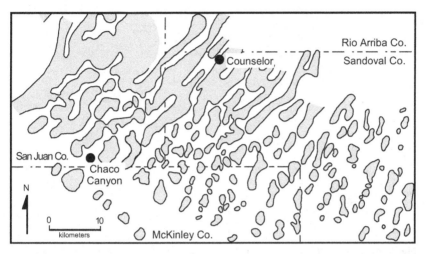

Fig. 6.19. Orientation and distribution of topographic ridges at the top of the Entrada Sandstone in northwestern New Mexico. The ridges represent eolian sand dunes, modified by wave-cutting and by sediment redistribution as they were buried beneath the overlying Todilto Formation. Present relief on the ridges is more than 30 m. From Kirkland *et al.* (1995).

dunes. To the west, the Entrada grades into tidal-flat red beds (Fig. 6.20). Deposition of the Entrada Formation was terminated by marine transgression. As the deepening waters of the Sundance Sea (on the Colorado Plateau also referred to as the Curtis–Summerville Sea) slowly inundated and eventually drowned the Entrada dune field, parts of the Entrada were deposited under water. Intertidal and deeper-water deposits include thin limestone and shale beds. Coquina beds may represent lag deposits formed in the wake of storms, and megaripples probably indicate tidal currents. The Entrada erg was located between 15° and 20°N latitude. Interpretation of the paleoenvironment (Fig. 6.21), including wind directions, suggests that the erg was deposited under prevailing trade winds in a hot, arid climate that from time to time was interrupted by typhoons (Kocurek, 1981).

In southern Colorado, northeastern Arizona, and northern New Mexico, marine inundation produced a shallow body of water, probably not more than about 90 m deep, with only a restricted or intermittent connection to the Sundance Sea, itself a shallow arm of the global ocean (Fig. 6.22).

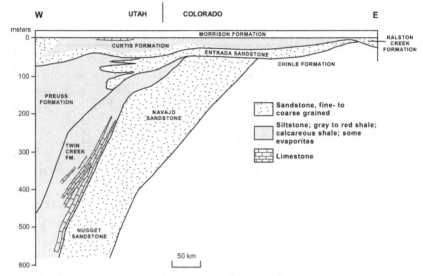

Fig. 6.20. The relationship of the Entrada erg to marine rocks of the Cordilleran miogeocline, indicated by the thickening of marine strata (especially the Preuss Formation) westward across Utah, is shown in this cross section across part of northeastern Utah and northwestern Colorado. Modified from Kocurek (1981). This figure emphasizes the great variability of the Jurassic section.

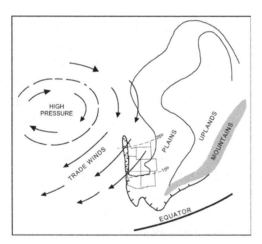

Fig. 6.21. Paleoenvironment of the Entrada Formation. At the time of the Entrada erg, the present Southwest was situated 15°–25° north of the equator. By analogy to modern meteorological conditions, an atmospheric high-pressure cell may have been located over the eastern Pacific to the northwest. Rotating clockwise, it may have created a generally southwest-directed trade-wind pattern of airflow over the continental margin. Modified from Kocurek (1981).

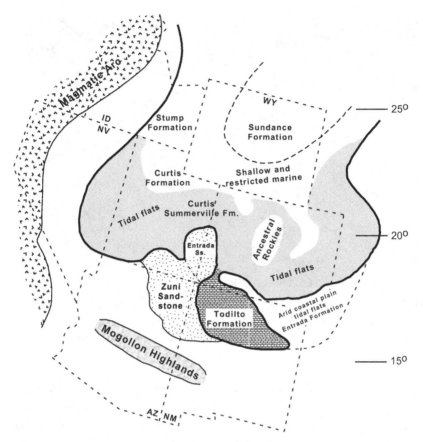

Fig. 6.22. Middle Jurassic paleogeography of the southern margin of the Sundance seaway and Todilto salina. Associated tidal flats and sabkha depositional systems are shown by gray pattern. Coastal ergs are represented by the Zuni and Entrada Sandstones (Ss.). From Anderson and Lucas (1996).

In the arid, Middle Jurassic climate, the body of water became evaporitic and hypersaline: a 'salina.' Marine water from the Sundance Sea mixed with fresh water from streams. The salinity of the water was influenced by seasonal influx of fresh water from streams and rainfall, by periods of drought, and by intermittent connections to the Sundance Sea. In this highly saline environment, first calcium carbonate was precipitated, forming a finely laminated limestone unit in which calcite laminae are separated by laminae of silt and organic material. Eventually, the basin became completely isolated from the Sundance Sea. As the salina became more saline,

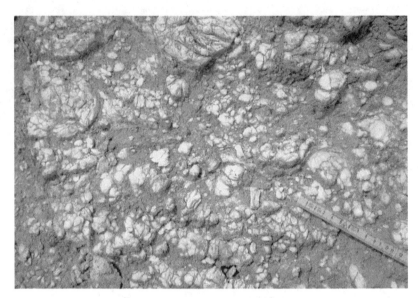

Fig. 6.23. Nodular gypsum of the Todilto Formation (Tonque Arroyo Member) (Lucas and Anderson, 1996), formed in a shallow, semi-isolated salina connected to the Sundance interior sea. Location is Kitchen Mesa, near Ghost Ranch, New Mexico. Scale is in cm.

gypsum was deposited, forming nodular and massive deposits. Together, these deposits, typically up to 37 m thick, make up the Todilto Formation (Fig. 6.23) (Armstrong, 1995; Kirkland *et al.*, 1995). Altogether, the Todilto salina probably lasted only a brief 30,000 to 100,000 years, representing only a very ephemeral marine deposit (Kirkland *et al.*, 1995). A modern analogue to the Todilto basin might be the 5000-year old MacLeod evaporite basin on the semiarid northwest coast of Australia. The narrow (*c.* 20 km × 110 km) MacLeod basin is separated from the Indian Ocean by a permeable bedrock barrier. Inflow of seawater by seepage through the barrier presently sustains a large-scale evaporite system in the basin, in which carbonate, gypsum, halite, and other salts are being deposited (Logan, 1987).

6.7 Return of continental conditions

Eventually, the Sundance Sea retreated and the Todilto basin began to fill with continental sediments. Conglomerate and sandstone deposited by streams with headwaters in the Sevier and Mogollon highlands built alluvial complexes, which prograded northeastward and eastward across more distal mudflats. The formerly lofty Uncompahgre uplift was by this time reduced to mere hills, contributing little to the basin fill. These sediments collectively form the lower Morrison Formation and equivalent units (Figs. 6.24, 6.25) (Peterson, 1988). The Morrison represents the first

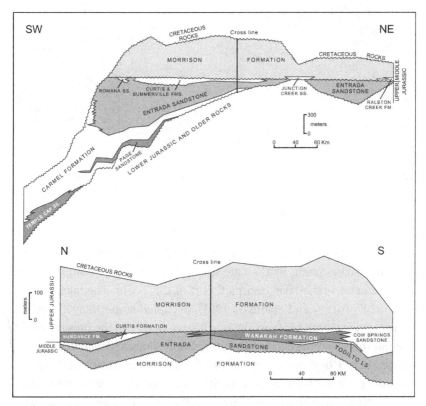

Fig. 6.24. Exposed Middle and Upper Jurassic rocks differ greatly in lithology across the Colorado Plateau and adjacent areas, as shown in these stratigraphic sections. The different lithologies reflect the wide range in paleogeographic conditions during deposition of these rocks, as well as subsequent uplift and erosion prior to deposition of Cretaceous strata. Upper section is NE–SW section from Denver, Colorado, to southwestern Utah. To the west, Jurassic units are abruptly truncated beneath Cretaceous units. Lower section is N–S along the Utah–Colorado and Arizona–New Mexico state border. Modified from Peterson (1988). Compare this section with Fig. 6.20, located farther to the west.

Fig. 6.25. The Brushy Basin Member of the Morrison Formation, seen in this photograph, is bentonitic, which gives rise to the whitish appearance. Location is Hall's Creek Overlook Road, south of Capital Reef National Park, Utah.

large-scale influx of sediments from the Cordilleran arc onto the Colorado Plateau (Blakey, 1996). By the Late Jurassic, a large playa complex, with an ephemeral saline lake, occupied part of this basin. The lake, named T'oo'dichi' (Turner and Fishman, 1991), occupied some 150,000 km^2 on the present eastern Colorado Plateau. It stretched from the present site of Albuquerque, New Mexico, to Grand Junction, Colorado, a distance of some 470 km. Lake T'oo'dichi' probably contained water only during infrequent and brief periods of intense storms, and was never deep. Between storms, the lake evaporated to dryness. Water level fluctuated greatly, depending on the frequency and intensity of storms. Only rarely did the lake bed fill with water. Typically, ephemeral streams from the basin margins traversed far out onto the dry lake bed, leading to a complex interbedding of fluvial and lacustrine sediments. Voluminous silicic ash, derived from volcanic centers in the arc to the west and southwest, was carried by prevailing westerly winds into the depositional basin, where it was washed and blown into the lake basin, forming claystone layers. In pore waters in the lacustrine sediments the ash altered during diagenesis to a variety of minerals, including

Fig. 6.26. Beds of conglomerate in the Upper Jurassic Broken Jug Formation. Clasts comprise approximately 80% limestone and 20% chert. In this area, conglomerate beds are strongly deformed. Limestone clasts are stretched, and some are deformed around or against chert clasts. Scale is in cm. Location is Broken Jug Canyon in the Little Hatchet Mountains of southwestern New Mexico (Lawton *et al.*, 2000).

smectite, clinoptilolite, analcime, potassium feldspar, and albite (Turner and Fishman, 1991). The presence of Lake T'oo'dichi' is interpreted as further evidence that the Jurassic climate was arid to semiarid. The Morrison Formation is rich in the remains of large, herbivorous dinosaurs, probably attracted to vegetation along the shore of the ephemeral water bodies. Lake T'oo'dichi' is the largest alkaline, saline lake known in the geologic record.

In contrast to the terrestrial setting described above, a different scenario played out along the southwestern margin of Laurentia. Here, Late Jurassic to Early Cretaceous rocks record a marine incursion, as a continental rift (the 'Mexican Borderland rift') formed behind (northeastward of) the Cordilleran volcanic arc. The Mexican Borderland rift was part of the larger Chihuahuan trough to the southeast, which itself was connected to the growing Gulf of Mexico that formed as Pangea broke apart and South America pulled away from Laurentia. Late Jurassic strata in south-central Arizona, at the northwestern tip of the continental rift, comprise a sequence of coarse conglomerates (the Glance Conglomerate of the Huachuca Mountains region). In southwestern New Mexico (Little Hatchet Mountains), strata approximately equivalent in age (the Broken Jug Formation) (Fig. 6.26) from the axis of the rift basin consist of silty marine carbonate and conglomerate (Lucas *et al.*, 2001). By the Early Cretaceous, this epicontinental sea reached south-central Arizona (Bilodeau, 1982; Lawton and McMillan, 1999; Lucas and Lawton, 2000).

Over most of the Southwest, however, continental conditions continued into the Cretaceous. The stage was now set for the last and greatest pulse of the Cordilleran orogeny, the Laramide.

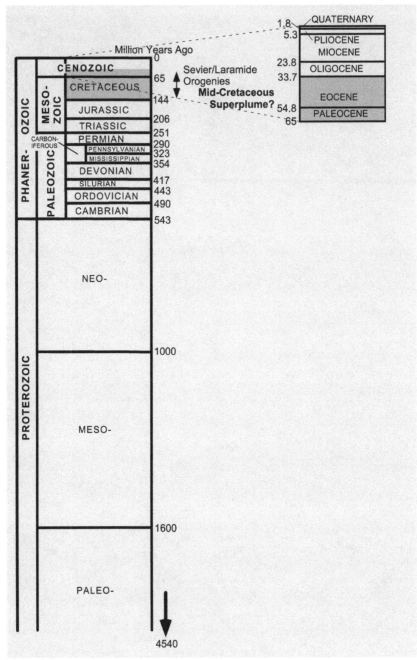

Geologic time scale. For explanation see p. 3.

Western orogeny

Cretaceous through Eocene

Tyrannosaurus rex prowled the margins of the Cretaceous Western Interior seaway. This cast of a *Tyrannosaurus rex* footprint from the Cretaceous–Tertiary Raton Formation near Cimarron, New Mexico, is approximately 76 cm from the tip of the longest toe to the end of the heel.

7.1 Active margin

Throughout the Cretaceous and into the Cenozoic Era, eastward subduction of the Farallon plate beneath continental crust of the North American plate proceeded seemingly unabated. Convergence resulted in the formation of an Andean-type margin, i.e. a subduction margin with a narrow magmatic arc developed directly on continental crust. Much of the Sierra Nevada and the Southern California/Baja batholiths, now the backbone of California and Baja California (Mexico), were emplaced at this time. These mountain ranges consist of numerous, overlapping plutons emplaced into the middle and upper crust beneath magmatic-arc volcanoes; most of the overlying volcanoes have long since been eroded away. The driving forces for orogenic pulses, characterized by crustal thickening and magmatism, were probably the result of coupling between the leading edge of the continent and east-subducting oceanic lithosphere of the Farallon plate. Docking and overthrusting of rafted terranes played only a minor role (Burchfiel *et al.*, 1992).

Compressional deformation occurred continuously throughout the Cretaceous and into the early Tertiary, with formation of uplifts and of sedimentary basins. In the area of the present Great Basin, the style of deformation was 'thin-skinned,' beginning earlier (see below) compared with regions to the east. This deformation, named the Sevier orogeny for the Sevier desert of Utah, is recognized mainly in Utah and Nevada, but extends from Arizona to British Columbia and Alberta. On the craton to the east, deformation resulted in basement-cored uplifts, beginning later than deformation to the west. This phase of deformation is the classic Laramide orogeny. By the Late Cretaceous, compressional deformation simultaneously affected an area stretching from the continental margin in California and Nevada eastward to Colorado and New Mexico. The remaining sections of this chapter, which discuss the somewhat arbitrarily separated phases of this basically continuous orogenic event, involve considerable overlap in time.

Deposition of continental sediments continued with little change from the Late Jurassic into the Early Cretaceous. On various parts of the Colorado Plateau, up to 300 m of fluvially deposited mudstone, sandstone, and conglomerate, laid down on lowlands and in occasional lakes, became the Cedar Mountain Formation. These sediments were deposited in

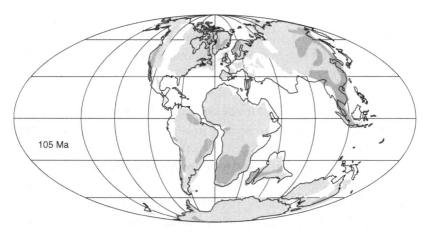

Fig. 7.1. Global paleogeographic map showing the relative locations of the continents during the Middle Cretaceous period, approximately 105 Ma. Stippled pattern indicates land areas, with darker pattern indicating higher elevation. Much of North America was covered by the Western Interior seaway. Modified from Smith *et al.* (1994).

a foreland basin formed in response to impending orogeny from the west (see below) (Lucas and Anderson, 1997).

The Cretaceous also saw another sea-level highstand, the Zuni (see Box 3.3), from about 100 to 80 Ma. The exact cause of the highstand remains unknown, but it may be related to the opening of the Atlantic Ocean following the breakup of Pangea (Heller *et al.*, 1996). North America continued its northward drift (Fig. 7.1), and by the Late Cretaceous may actually have been slightly north of its present latitude (Smith *et al.*, 1994).

7.2 The severe Sevier

Although part of a protracted and nearly continuous period of compressional tectonism, the Sevier orogeny was qualitatively very different from, and more severe than, the preceding orogenies of the Paleozoic and Mesozoic eras. The Sevier orogeny eventually affected regions more than 1000 km eastward through the Paleozoic miogeocline and well into the North American craton. Sevier deformation is characterized by often-intricate compressional folds and, most prominently, by imbricated, low-angle thrust faults developed mainly in miogeoclinal and cratonic sedimentary rocks. In fact, the fold-and-thrust belt of the western USA and Canada (Fig. 7.2) is one of the most important thrust belts in the world, for its scale as well as for the understanding that has been gained from the many studies of the belt (DeCelles *et al.*, 1995).

The timing of inception of the Sevier orogen is not at all clear, and estimates range from the latest Jurassic to late Early Cretaceous. Certainly, major deformation was in progress approximately 100 Ma in parts of Utah and Wyoming. In the Sevier Desert region of Utah, the Sevier belt comprises four major thrust systems, which together accommodate at least 120 km of east–west crustal shortening (DeCelles *et al.*, 1995). Overall, thrusting progressed from west to east, probably ending by about 50 Ma (DeCelles and Mitra, 1995).

Box 7.1 Isostatic flexural subsidence. Crustal basins typically develop in the forelands (i.e. in *front* of, as viewed from the stable craton) of thrust-sheet orogens, such as the Sevier, through a mechanism described as 'isostatic flexural subsidence.' The term describes the process by which crust that has been loaded (hence, thickened) by any process, such as by the piling up of thrust sheets, becomes out of *isostatic* equilibrium. It regains equilibrium by *subsiding* deeper into the mantle. Because the upper crust is relatively cool and therefore strong, the adjacent, unthickened crust is *flexed* downward for some distance outward from the orogen, forming an asymmetrical basin in the foreland adjacent to the thrust belt.

Foreland basins contain great thicknesses of sediments, derived mainly from the adjacent orogen. The sediments also contribute to the load on the crust. Once the stratigraphy and ages of sediments in the basin are understood, the tectonic history of the basin, hence of the orogen, can be deciphered by 'backstripping' of sediments, i.e. by systematically stripping off sediments of progressively greater age. Backstripping reveals the volume and shape of the basin at various times during its development (Fig. 7.2).

Although compressional deformation had little direct effect in most of the Southwest, an extensive **foreland basin** developed (see Box 7.1) as the craton subsided in response to the load superimposed by the eastward-advancing wedge of thrust sheets (Fig. 7.2). The advance of thrust faulting during the Cretaceous (culminating in the Sevier fold-and-thrust belt) resulted in disruption of the original western edge of the foreland basin

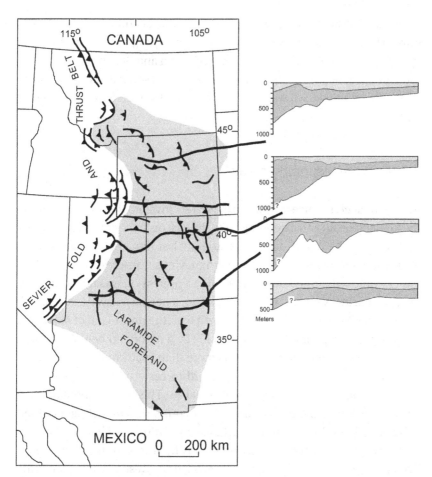

Fig. 7.2. Generalized tectonic map showing locations of major thrust faults of the Sevier thrust-and-fold belt and of the Laramide (or Rocky Mountain) foreland (shaded). The foreland is the area in front of the Sevier orogen that was affected mainly by basement-cored, high-angle faults. The four cross sections are subsidence profiles (not geologic units) across the foreland basin adjacent to the Sevier fold-and-thrust belt. The profiles are derived by stripping of sediments to derive basin shape at time intervals of interest (see Box 7.1). Light shading indicates magnitude of subsidence from 97 (94 on lower two profiles) to 90 Ma; dark shading is subsidence from 90 to 80 (83 on upper two profiles) Ma. Vertical scale is tectonic subsidence in meters. Locations of profiles are shown by dark lines. From DeCelles and Mitra (1995) and Pang and Nummedal (1995).

as basin deposits were caught up in thrust sheets and transported eastward. The depositional axis of the basin shifted progressively eastward as thrust sheets encroached from the west (Cowan and Bruhn, 1992). During the Late Jurassic, the asymmetrical basin, deepest adjacent to the uplift where the applied load was greatest, extended some 450 km to the east (Fig. 7.2).

7.3 Western Interior seaway

In the Cretaceous, once again, marine waters invaded the craton, flooding the foreland basin and extending far to the east onto the craton. This transgression was the Zuni, and represented the last time that the seas would invade the Southwest. In contrast with previous transgressions, the shallow cratonic sea formed during the Zuni transgression was not connected to open marine waters to the west. Instead, a narrow 'seaway' developed, connecting the Arctic Ocean to the north with the proto-Gulf of Mexico to the south and separated from the Pacific by a mountain belt. At its maximum development, the seaway was more than 1000 km wide, stretching from western Arizona and Utah eastward to eastern Missouri and Iowa (Fig. 7.3). Water depths ranged up to 400 m. This epicontinental sea is named the 'Western Interior seaway.' From numerical modeling (Ericksen and Slingerland, 1990), it is inferred that circulation (and hence sediment transport) in the seaway was driven mainly by storms (rather than tides). Typical winter storms crossed the seaway from west to east, generating longshore currents flowing mainly to the south. Extreme storms could have produced currents with velocities of up to 0.8 m/s over the shelves, and waves up to 9 m high.

What caused this narrow inland sea to develop? Why did it divide North America into two land areas, in contrast with previous flooding events, which simply flooded the margins of the craton? Part of the explanation, of course, was the Zuni marine transgression, during which sea level rose 175–250 m above the present level (Algeo and Seslavinsky, 1995). However, a more important part of the explanation is associated with both static and dynamic mechanisms related to the deformation of the western margin of the continent, arising from ongoing subduction of the Farallon plate beneath the North American. The Sevier orogeny thickened the crust,

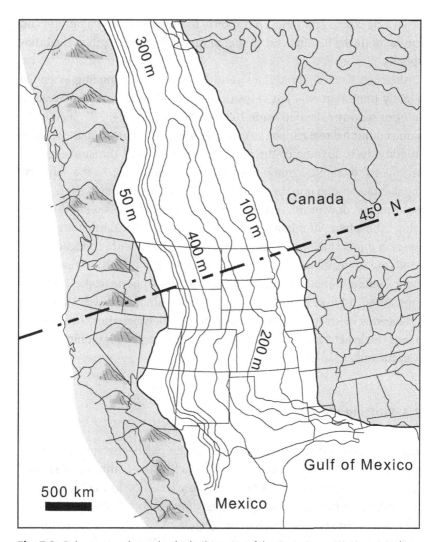

Fig. 7.3. Paleogeography and paleobathymetry of the Cretaceous Western Interior seaway near its maximum development about 90 Ma (Early Turonian). From Ericksen and Slingerland (1990).

which subsided under the accumulated load of the thrust sheets, creating a foreland basin extending at least 400 km eastward from the fold-and-thrust belt onto the craton (Fig. 7.2). The basin was asymmetrical, with its deepest part closest to the orogen (Chapter 5). Subsidence rates were as high as 64 m/Myr for a portion of the Late Cretaceous, but differed greatly at various places and times in response to differences in magnitude and timing of thrust loading and to spatial variations in strength of the lithosphere.

Although flexural subsidence resulting from thrust loading extended slightly more than 400 km eastward from the Sevier front, the Western Interior seaway extended some 1600 km to the east (Fig. 7.3). Seemingly, some additional mechanism to explain the crustal depression is required. In addition to loading of the crust by superimposed thrust sheets and sediments, dynamic loading from below also occurred. As the subducting slab descended into the mantle, the asthenosphere that overlay it also was dragged downward, resulting in topographic depression in the overlying lithosphere. At relatively rapid subduction rates, particularly when the slab dip is shallow, the wavelength of the subsidence induced in the overlying lithosphere can be significantly greater than that induced by static loading alone. A combination of supralithospheric (static) and sublithospheric (dynamic) loads is thought to explain flexural warping of the Cretaceous Western Interior seaway. Interplay between these two mechanisms can lead to very complex sedimentologic and stratigraphic patterns (Catuneanu et al., 1997).

7.4 Shifting shores

The Western Interior seaway dominated the paleogeography and sedimentation of the Late Cretaceous over much of the Southwest (Fig. 7.4). Its relative sea-level changes are intimately recorded in the rock record, well displayed in the Four Corners area of the Colorado Plateau. Here, at least two major and numerous minor regressions, stretched out over some 25 Myr, are preserved in the stratigraphic record. The position of the shoreline was marked by sand, as beach sand and sand bars migrated back and forth in response to changes in sea level. The various and complicated advances and withdrawals of the sea gave rise to a highly sinuous and nearly continuous (through time) sand-dominated facies that shifted its location back and forth through time with advances and retreats of the sea (Fig. 7.5). At any given place, transgression resulted in an upward progression in lithologies from strand-line sandstone to marine shale, and the reverse for regression. The sand unit separates continental deposits, typically

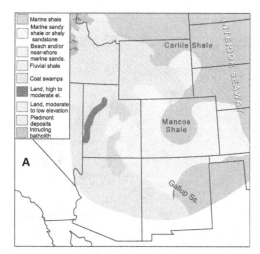

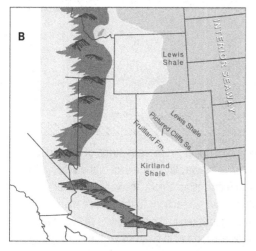

Fig. 7.4. Paleogeographic maps of the western margin of the Cretaceous Western Interior seaway of Late Cretaceous time. Selected characteristic formations are identified. The relationships of different lithologies to sea level, and the changing environments of deposition through time, are indicated. (A) The seaway approximately 90 Ma, during a minor regression in the Mancos sea. (B) Approximately 74 Ma, during the final withdrawal of the Lewis sea. Times are indicated in Fig. 7.5. From McGookey *et al.* (1972). See also Plate 15.

coal-bearing, west of the strand deposits from shallow marine strata to the east, within the seaway. The sandstone formations corresponding to the sand-dominated facies have different names in different places, depending on where they were first recognized and described.

The first transgression of the inland seaway began with sand and related lithologies of the upper Cretaceous Dakota Formation, a fluvial, deltaic, and marginal-marine deposit (Fig. 7.6). Throughout the Four Corners area, the Cretaceous seaway spread slowly westward over the alluvial plain. From Colorado westward to central Utah the rate of migration of

SW   NE

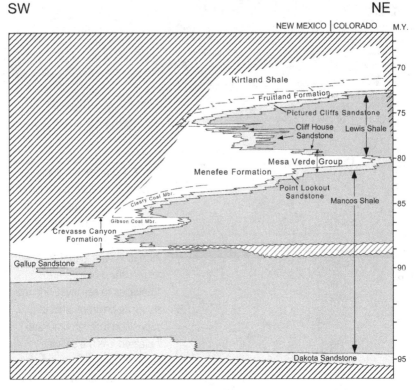

Fig. 7.5. Regional stratigraphic section of Upper Cretaceous strata of the San Juan Basin. The axis of the Cretaceous Interior seaway lay to the east of the area represented in this cross section. The relationship among the various littoral sandstone deposits (light gray) is shown by this section. To the east lie marine shale, siltstone, and limestone (medium gray) deposited in the seaway. To the west are non-marine deposits (white). Coal typically formed in swamps on the landward side of the beaches. Diagonally ruled pattern indicates areas of section removed by erosion. Simplified from Molenaar (1983).

the Dakota shoreline averaged a slow 0.15 m/Myr (Fillmore, 2000). Many **dinosaur trackways** (Box 7.2) are preserved in rocks of the Dakota Formation (Fig. 7.7).

Deep-water deposits of the seaway consist principally of dark, fossiliferous shale and mudstone, with lesser thickness of interbedded sandstone, limestone, and marlstone. These rocks form the Mancos Shale, which in southwestern Colorado ranges to over 700 m in thickness. The drab color of the Mancos, which contrasts sharply with the brightly colored strata that

Fig. 7.6. Burrowed zones of the crustacean *Thalassinoides* ornament bedding planes in the upper sandstone unit of the Dakota Sandstone (Grant and Owen, 1974). Location is Rim Vista, 4 km west-northwest of Ghost Ranch, New Mexico. Scale is approximately 13 cm in length.

Box 7.2 Dinosaur trackways. The footprints (tracks) left behind by dinosaurs and the trails that they traveled (trackways) allow us to glimpse attributes of dinosaurs that are not provided by their physical remains. They complement the skeletal record in very significant ways, including demonstrating presence and activity where skeletal remains are not known. Tracks and trackways show step and stride patterns, revealing information about locomotion, posture, and gait. Trackways can tell about social behavior, such as how many animals were active in an area, what mixture of large and small animals existed, and something of their activities. Trackways provide evidence that some dinosaurs were gregarious, traveling together in large herds. They offer paleoecological insight into the relationship with habitat, which is relevant to an understanding of metabolism and biogeography. Tracksites (areas of sedimentary strata preserving tracks) in the Dakota Group east of the Front Range in Colorado, Oklahoma, and New Mexico provide evidence for a 'dinosaur freeway' along the sandy and muddy coastal plain of the Western Interior seaway (Fig. 7.7). Altogether, this **megatracksite** may comprise an area of more than 75,000 square kilometers. New tracksites are discovered every year, and tracking dinosaurs has become a vital part of dinosaur research (Lockley *et al.*, no date; Lockley, 1998).

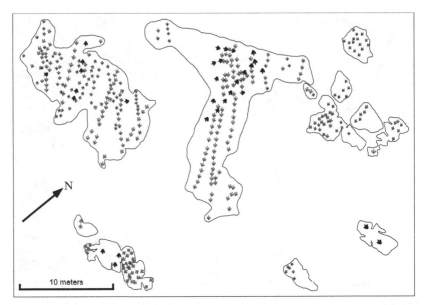

Fig. 7.7. Map of the dinosaur tracksite at Mosquero Creek, New Mexico, provides evidence for a herd of at least 55 dinosaurs walking northward along the shore of the Western Interior seaway during deposition of the Dakota Sandstone. Animals were mostly juvenile iguanodontid dinosaurs. The large number of tracksites along the shore of the seaway, and the large amount of dinosaur 'traffic' at these sites, suggest that this site was part of a major thoroughfare, a 'dinosaur freeway.' Lines surrounding groups of tracks designate areas of exposed strata. Modified from Lockley *et al.* (no date).

characterize so much of the Southwest, indicates relatively low-oxygen conditions during deposition and burial. Numerous layers of bentonite (altered volcanic ash) occur throughout the Mancos Shale. The bentonite layers, which are essentially instantaneous and widespread time layers, enable correlation of parts of the Mancos Shale in the Four Corners area with units as far east as eastern Colorado and western Kansas (Fig. 7.4). The excellent stratigraphic and age control provided by fossils and key marker beds allows geologists to determine that sedimentation rates were as high as 195 m/Myr in parts of the Interior sea in southwestern Colorado (Fig. 7.8) (Leckie *et al.*, 1997).

Gradually and haltingly, beginning some five or so Myr after the first Late Cretaceous transgression of the seaway, the marine waters slowly

AGE (MA)

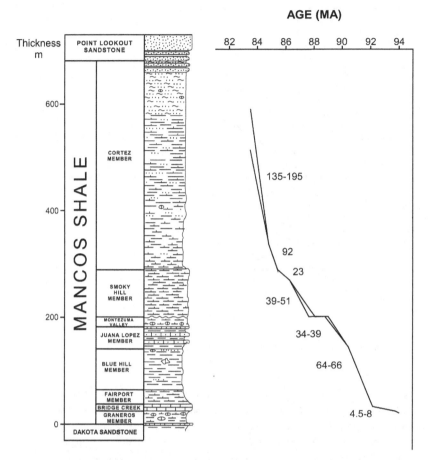

Fig. 7.8. Estimates of sedimentation rates in a reference section through the Mancos Shale near Mesa Verde National Park, Colorado. Lower rates in some units may, in part, indicate missing record. Separate curves for several horizons indicate uncertainties in rates. From Leckie *et al.* (1997).

receded. Sand deposits from the margin of the receding sea built out over deeper-marine mud. The shore-edge sand now forms the Gallup and Point Lookout Sandstones, which form time-transgressive horizons capping the Mancos Shale (Fig. 7.5). The Point Lookout Sandstone, a strand line and deltaic deposit (Fig. 7.9), has been particularly well studied because it is an important reservoir for oil, gas, and potable water in the San Juan basin of northwestern New Mexico and southern Colorado. In detail, it is a complex unit, comprising linear-strandplain, barrier-island,

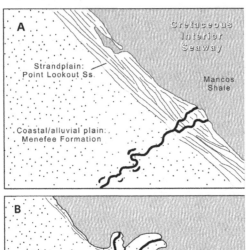

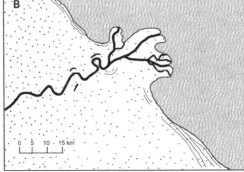

Fig. 7.9. Two models for the depositional environment of the Point Lookout Sandstone in New Mexico and Colorado, deposited along the western margin of the Western Interior sea. (A) A strandplain-barrier island model; (B) a deltaic model. Both are probably valid for different geographical areas. Seaward of the coastline, black mud (the present Mancos Shale) was deposited in the marine waters. On the alluvial plain landward of the coast, fine sand, silt, and mud of the terrestrial Menefee Formation was laid down. The entire 'package' of formations migrated eastward through time as the shoreline advanced. Modified from Wright-Dunbar *et al.* (1992).

intertidal (e.g. estuary), back-beach, and shallow-shelf deposits. Landward of the strandline, sand, silt, and gray, carbonaceous mud was deposited by streams flowing toward the coast across an alluvial plain. Behind the receding strand, marshes and lagoons formed, in which peat accumulated. These deposits make up the Menefee Formation, which interfingers with and overlies the Point Lookout Sandstone.

Regression was not smooth and continuous, but rather punctuated by numerous reversals in relative sea level. As marine waters, in bays, lagoons, and channeled estuaries, spread onto the land, dispersing and eroding the beach dunes, marine muds accumulated over sands of the shallow inner shelf and beach (Figs. 7.10, 7.11, 7.12). Then, relative sea level again fell and the coastline prograded over marine facies. Sand was again transported into the open marine environment as shoreface deposits spread over marine mud. The result of the minor transgressions in an overall regressive shoreline is cyclic alternations of marine sandstone and shale (Fig. 7.10).

For a second time, the western margin of the vast Interior seaway expanded over the continent, although this phase was more areally restricted. Again, well-sorted sand deposits from beaches, bars, and coastal dunes, and nearshore marine sand and mud spread over the continental sediments. The sand deposits of this second transgression are known as the Cliff House Sandstone. Together with the underlying Point Lookout Sandstone and continental Menefee Formation (Fig. 7.13) between, they form a triad of units referred to as the Mesa Verde Group because they cap the well known Mesa Verde of southern Colorado. The deposits of this sea form the Lewis Shale, a gray to black shale formation with interbedded sandstone, limey units, and bentonite beds very similar to the Mancos Shale. In the Four Corners area, the Lewis shale attains thicknesses of over 500 m.

The second retreat of the sea was marked by sand deposits of the Pictured Cliffs Sandstone. This sandstone is similar in every way, including environment of deposition, to the Point Lookout Sandstone that was formed by the regression of the Mancos sea. In turn, the Pictured Cliffs Sandstone was overlapped by thick deposits of fluvial, deltaic, and paludal sediments, such as the Fruitland and Kirtland Formations (Fig. 7.14) that accumulated on alluvial plains and coastal marshes. The Fruitland Formation contains coal beds, in places more than 10 m thick (Bland, 1992), and is the major coal-producing unit in the Four Corners area.

The retreat of the Lewis sea in the Late Cretaceous was the final withdrawal of the sea from the North American continent, and the beginning of the modern sea level 'lowstand.' More of the continent is exposed now than at most times during the Phanerozoic (Algeo and Seslavinsky, 1995). Subaerial conditions prevailed throughout the Southwest during the early Tertiary. Sediments consisted of conglomerate, sandstone, and mudstone derived from intermittently rising uplifts of the Sevier and Laramide orogens. The sediments were deposited on alluvial fans, and along streams and in swamps on alluvial plains in the intervening basins. In southern Colorado and northern New Mexico, dust fallout from the **K/T boundary**

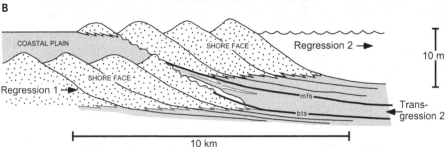

Fig. 7.10. (A) Point Lookout Sandstone (upper cliff) overlies the Mancos Shale in this cliff exposure at Bosque Grande Mesa, 38 km south-southwest of Cuba, New Mexico. The Point Lookout Sandstone is a shore face (beach) sand deposited on marine mud of the Mancos Shale as waters of the Cretaceous Western Interior seaway slowly withdrew from the land. Overall regression was interrupted by several minor marine transgressions. Three of these, marked T_1 through T_3, are clearly visible in this exposure. The transgressive units comprise 'ramps' (because they were deposited on gently seaward-sloping surfaces) of marine mudstone deposited over shoreface and shallow inner shelf sandstone (Lucas *et al.*, 1992). (B) Schematic model illustrates formation of a thin transgressive depositional ramp (Transgression 2), consisting dominantly of mudrock (shaded pattern), between regressive shoreface deposits (Regression 1, 2), comprising beach sand (coarse stipple). Transgressive deposits are progressively finer-grained upward from basal sandstone (fine-stippled) overlying the basal transgressive surface (bts) to silt and clay-rich mudrock at the maximum flooding level (mfs). See Box 4.1. Modified from Wright-Dunbar (1992).

Fig. 7.11. Soft-sediment deformation in the lower part of the Point Lookout Sandstone, seen here near the entrance to Horse Canyon, Durango, Colorado. Basal sediments of the Point Lookout Sandstone were deformed owing to rapid loading of sand onto the contact with mud of the underlying Mancos Shale. The zone of soft-sediment deformation is approximately 3 m thick. The overlying Point Lookout Sandstone is evenly bedded and undisturbed. The fine-grained, unit below is the Mancos Shale. Scale is 1 m.

Fig. 7.12. Transgressive marine mudstone unit T_2 (Fig. 7.10) of Wright-Dunbar (1992) from the gradational contact zone between the Mancos Shale and the overlying Point Lookout Sandstone is typified by yellowish to reddish staining and casts of vertical, smooth-sided burrows of marine organisms. Same location as Fig. 7.10. Left black–white bar is in cm.

Fig. 7.13. As the Mancos Shale was deposited in the shallow Cretaceous seaway, the Menefee Formation, seen in this photograph west of Durango, Colorado, was laid down by streams flowing across the coastal plain. The Menefee consists of sand, silt, and gray, carbonaceous mud, forming rather drab-colored outcrops. It contains significant deposits of coal, which accumulated as peat in marshes and lagoons behind the receding strand. The Menefee Formation interfingers with and overlies the Point Lookout Formation.

event (Fig. 7.15; Box 7.3) was preserved in coal swamps (Pillmore and Flores, 1990). Because the Late Cretaceous (Fig. 7.16) and Early Tertiary sediments reflected the complex paleogeography of the newly developed Cordillera, the character of the sediments differed greatly over short distances and typically changed dramatically over short time intervals, leading to a plethora of local formation names.

7.5 Laramide orogeny

The Laramide orogeny was arguably the greatest mountain-building episode to affect the western USA. Rather than being a separate and distinct orogenic event, the Laramide was actually a continuation of the Sevier orogeny. Extending from latest Cretaceous to the Middle Eocene (approximately 80–40 Ma), the Laramide overlapped in time with the later part of the Sevier event (Section 7.2). However, it affected a different part of North America and, therefore, entailed a very different style of deformation.

A

B

Fig. 7.14. (A). The Naashoibito Member of the Kirtland Formation contains numerous petrified logs, such as the one shown in this photograph. The location is the De-Na-Zin Wilderness, south of Farmington, New Mexico. Scale is in cm. (B) Details of knots and even of the grain structure of this 74 m.y.-old tree in the Kirtland Formation are well preserved. Location is same as (A); scale is in cm. See also Plate 16.

Fig. 7.15. Cretaceous–Tertiary boundary layers are preserved in coal-bearing sediments of the Raton Formation in northern New Mexico and southern Colorado. The prominent white layer is the claystone layer, composed of altered glassy ejecta launched ballistically from the Chicxulub impact site. It represents a bad day for the Earth. Above it (immediately below the scale) is the impact (or 'fireball') layer, formed as shocked rocks from the impact area, lofted into the atmosphere by a fireball, settled out over the subsequent weeks and months. This layer contains grains of shocked quartz and feldspar, and high concentrations of iridium. Deformation of the claystone and impact layers is due to the weight of the overlying strata. The boundary layers overlie a layer of carbonaceous shale and are overlain by a bed of coal. See Box 7.3. Scale is in cm. Location is 4.6 km south of Starkville, Colorado. Reference: Pilmore and Flores (1987); Izett (1990). See also Plate 17.

Whereas the Sevier deformation was primarily characterized by thrusting and folding that was confined to Phanerozoic strata overlying crystalline basement rocks, the classical Laramide was characterized by uplifts of Precambrian crystalline rocks ('basement-cored' uplifts) along high-angle-reverse faults. And although the Sevier fold-and-thrust belt (the eastern limit of what is regarded as the Sevier orogeny) approximately coincides with the continent-ward edge of the miogeocline, the Laramide orogeny affected the craton as far east as the Black Hills of South Dakota, the Front Range of Colorado, and the Sangre de Cristo Mountains of New Mexico, 1000–1500 km from the plate margin in present coordinates.

Fig. 7.16. Boulder conglomerate of the McDermott Member of the Animas Formation. Clasts of andesite up to at least 90 cm in longest dimension characterize the McDermott at this location, 5.6 km south of Durango, Colorado. The McDermott Member is latest Cretaceous in age (Anderson *et al.*, 1997). Scale is 1 m.

Box 7.3 End-of-Cretaceous event. The extinction event at the end of the Cretaceous was the *second* greatest die-off in the history of the Earth, after that marking the end of the Paleozoic (Chapter 5). Its primary claim to fame was that it was marked by the demise of the dinosaurs, but its effect on life was far greater. Three quarters of marine species went extinct at the end of the Cretaceous (Jablonski and Raup, 1995), including 95% of free-floating foraminifers. Among terrestrial fauna, and to the extent that eastern Montana is representative of other regions of Earth, 10% of species from freshwater environments and 88% of species from land-dwelling assemblages became extinct (Sheehan and Fastovsky, 1992).

Although many geologists resisted the idea of an extraterrestrial cause for years after it was proposed by Alvarez *et al.* (1980), the evidence that this extinction event resulted from impact of a meteorite or comet now seems overwhelming. The first evidence for an extraterrestrial cause came from marine sediments now exposed on land, where a layer anomalously rich in the platinum-group element iridium (Ir) was detected. Ir is rare in terrestrial rocks but common in meteorites. Besides Ir, the Cretaceous–Tertiary (abbreviated K/T) boundary is marked by grains of shocked quartz, feldspar, and zircon; by relict tectites (glass); and by Ni-rich spinel and diamonds (Hough *et al.*, 1997). In many areas, the boundary is also marked by elemental sulfur, carbon, soot, and fullerenes (C_{60} and C_{70}) (Wolbach *et al.*, 1990; Heymann *et al.*, 1998). Some of the early supporting evidence for impact came from the Southwest. Near Raton, New Mexico, and Trinidad, Colorado, stratigraphic sections formed in and adjacent to coal swamps, in which deposition was continuous from the Cretaceous into the Tertiary, similarly contain a layer anomalously enriched in Ir. At these localities, the K/T boundary is characterized by a pair of thin claystone units (Fig. 7.15): (1) a lower bed, the K/T 'boundary claystone,' 1–2 cm thick, composed mainly of kaolinite with few or no shock-metamorphosed mineral grains, and (2) an overlying bed, the 'impact layer,' averaging 5 mm thick. Ir is most abundant (up to 15 ppb) in the impact layer, which is rich in shocked mineral grains, especially quartz (Izett, 1990). Interestingly, the couplet of claystone layers is overlain by coal, in contrast with the extinction that marks the end of the Paleozoic Era (Chapter 5). The Raton–Trinidad sections were important in providing evidence that the event was global and not merely restricted to oceans.

However, the most compelling evidence, the 'smoking gun,' so to speak, is the presence of a large impact crater near Chicxulub, on the Yucatán Peninsula of Mexico. The crater, which lies partly offshore, is buried by about a kilometer of post-Cretaceous strata. It has been dated at 65 Ma, precisely the age of the K/T boundary within analytical error (Swisher *et al.*, 1992). The size of the transient crater, the hole initially blasted in the crust by the impact, was about 100 km in diameter and 30 km in depth, but subsequent collapse and uplifts formed a multiringed crater with a bounding, annular fault scarp, similar to craters observed on Venus and the Moon, with a total diameter of 170–195 km (Morgan *et al.*, 1997). The collision was so severe that it deformed all levels of the crust and the uppermost few kilometers of the mantle as well (Snyder and Hobbs, 1999; Christeson *et al.*, 2001). The Chicxulub impactor must have been 12–15 km in diameter, traveling at 15–20 km/s. Objects of this size or larger strike the Earth about every 100 million years.

The boundary claystone (or 'melt ejecta layer') is unique to sites in western North America and the Gulf of Mexico. It is now interpreted to represent altered glassy ejecta, launched as a curtain of melt and shocked rocks. In the Southwest, particles comprising this layer likely were emplaced within 15 minutes or so of impact (Alvarez *et al.*, 1995). The thorough alteration of glass comprising the bulk of this layer is attributed to the effect of acids produced by the impact and ensuing environmental consequences (Retallack, 1996). In contrast, the impact (or 'fireball') layer is similar to that in exposures of the K/T boundary at many other sites outside of North America. It represents shocked rocks from the impact area launched in a fireball of CO_2 and steam, and would have been deposited considerably after the glassy ejecta layer (Alvarez *et al.*, 1995). From measurements of [3]He in exposures of the impact layer in deep-marine sections now exposed on land in Italy, it has been determined that the clay layer was deposited over a time interval of about 10 ka, the time required to restore food chains and repair ecosystems (Mukhopadhyay *et al.*, 2001).

The asymmetry of the Chicxulub impact crater, together with evidence from the distribution of shock-metamorphosed minerals, restriction of the boundary claystone to western North America, and palynofloral extinction and survival patterns, are interpreted to indicate that the impacting bolide followed a low-angle trajectory (20–30° above the horizon) from the southeast (Schultz and D'Hondt, 1996; Snyder and Hobbs, 1999).

One question remains unanswered. Unless actually struck by the meteorite or close enough to it to be destroyed by the blast (estimated at equivalent to 100,000 megatons of energy) (Alvarez *et al.*, 1995), what was the actual mechanism of destruction? Debris thrown into the atmosphere by the impact, along with carbon, soot, and ash from resulting fires, probably darkened the skies for weeks or months following the impact, but not more than for a few years at the longest (Mukhopadhyay *et al.*, 2001). It has been estimated that the average global surface temperature decreased by from 12 to 19°C for up to 9.5 years, producing long-term cold conditions (but not freezing, because global surface temperatures in the Late Cretaceous were 18–20°C warmer than at present) (Gupta *et al.*, 2001). From the amount of polycyclic aromatic hydrocarbons of pyrosynthetic origin found at the boundary, it has been estimated that 18–24% of the Earth's above-ground biomass may have burned following the impact (Arinobu *et al.*, 1999). Wildfires were global in extent (Vajda *et al.*, 2001). Debris rained down onto the Earth's surface for months afterward. Over 1 trillion tons of CO_2 and S_2 were released (Cygan *et al.*, 1996), and precipitation following the event was probably highly acidified. A small but significant change in the $^{87}Sr/^{86}Sr$ composition of sea water following the K/T boundary event may be related to enhanced weathering of continental rocks due to the effect of acid rain (MacLeod *et al.*, 2001). Altogether, the effects (probably greatest in western North America) constituted an unprecedented environmental disaster.

Finally, as satisfying as it may be to attribute so confidently a dramatic extinction event to a simple cause, mass extinctions may result from a variety of causes besides the impact of an extraterrestrial object. It seems likely that at least some extinction events may have occurred from complex causes intrinsic to the Earth's own workings (Hames *et al.*, 2000).

The effect of the Laramide orogeny on the Southwest was very uneven. On the present Colorado Plateau, it created gentle uplifts and shallow basins. Between the uplifts and basins, monoclinal flexures formed where sedimentary strata draped off the sides of the uplifts. Today, these

Fig. 7.17. The Raplee monocline, near Mexican Hat, Utah, resulted from crustal shortening during the Cretaceous–Tertiary Laramide orogeny. See also Plate 18.

eroded monoclines make spectacular linear hogbacks, such as the Raplee monocline near Mexican Hat, Utah (Fig. 7.17), Comb Ridge west of Bluff, Utah, and the Hogback monocline near Ship Rock, New Mexico, but are relatively minor structural features. Farther to the east, the effects of the orogeny were much more severe. Orogeny raised high mountain ranges, coinciding in part with the Front Range of Colorado, the Sangre de Cristo Mountains of southern Colorado and northern New Mexico, and other ranges of southern New Mexico (Fig. 7.18) (Chapin and Seager, 1975). In many cases, uplifts created in the Laramide orogeny coincided with those raised by the Ancestral Rocky Mountains orogeny of Pennsylvanian age (Chapter 5), but long since eroded away (Fig. 7.18). This coincidence of location seemingly indicates that the continental crust, once broken by orogeny, requires hundreds of millions of years to 'heal,' if it ever does. In southern New Mexico, northern Chihuahua, and west Texas, the Laramide developed a 'thin-skinned' fold-and-thrust belt, the Mexican fold belt, oriented in a northwest–southeast direction (Fig. 7.18). As discussed in the next chapter, this trend controlled later extensional basins.

Although it has always been clear that the Laramide represented major crustal shortening and uplift of the Cordillera, the cause or causes of

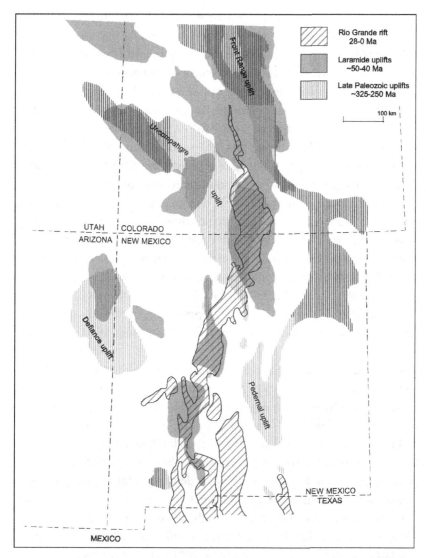

Fig. 7.18. Uplifts and structures of the Laramide orogeny (shaded pattern) in New Mexico and Colorado generally follow uplifts of the Ancestral Rocky Mountains of late Paleozoic age (vertically lined pattern). Similarly, the modern Rio Grande rift (diagonally lined pattern), described in the next chapter, generally follows uplifts of the Laramide orogeny. Modified from Chapin and Seager (1975).

the orogeny are very much under debate. Probably the most generally accepted explanation, which may be largely correct, is that the Laramide uplifts and basins are a direct response to rapid, northeast-directed convergence between the Farallon and North American plates. A relatively high convergence rate meant that the North American plate overrode the hot, buoyant Farallon plate faster than the Farallon plate was able to sink. As a result, it was subducted beneath the North American plate at a very shallow angle (5–10°). Along most parts of the subduction zone the magmatic arc disappeared altogether. Along other parts, the shallow dip increased the distance between the trench and the volcanic front, a broad volcanic belt. The composition of the volcanic belt tended to be more alkalic than typical of subduction-related magmatic zones (Lipman, 1980). Coupling between the two plates created compressional stresses that propagated upward through the continental lithosphere, creating crustal shortening and uplift, with intervening basins (Dickinson and Snyder, 1978; Burchfiel *et al.*, 1992; Yin and Ingersoll, 1997). Yet, the mantle root of the North American plate is known to have been preserved, because its geochemical imprint is clearly discerned in modern basaltic magma (Livaccari and Perry, 1993).

A modern analog for the Laramide orogeny may be portions of the Andes Mountains in Chile and northern Argentina (latitude 25–34°S), where subduction of the Nazca plate is very shallow (5–10°). Here, east of the magmatic arc lies a region characterized by a thin-skinned thrust belt, similar to the Sevier thrust belt in Utah, and a broad foreland region characterized by uplifts of crystalline basement (including the Pampeanas Ranges), analogous to the Laramide uplifts of New Mexico and Colorado. Despite these similarities, the uplifts of the Laramide in the Southwest developed greater structural relief, occurred farther from the plate margin, and evolved over a longer time span than has deformation in this part of the Andean margin (Jordan *et al.*, 1983; Kay and Mpodozis, 2001). The mechanisms of deformation and effects on the lower crust and lithospheric mantle are not clear. By the end of the Laramide orogeny in the Late Eocene, the western United States resembled a high, wide mountain belt, much like the Chilean and Argentinian Andes of today. The average elevation of much of the region was probably about 3 km (Wolfe *et al.*, 1997, 1998). Continental sediments of Eocene age were deposited across a landscape dominated by erosion of previously high Laramide uplifts.

Some of the mountainous terrane uplifted by the Laramide, such as the Front Range of Colorado and the southern Sangre de Cristo Range of northern New Mexico, remains high today (Kelley and Chapin, 1995), for the 40 or so Myr that has elapsed was insufficient to wear down these lofty ranges. Yet the mountains formed by the Laramide were not the ranges that we see today. In places, the Laramide ranges were diminished by subsequent collapse, such as along parts of the Rio Grande rift (Baldridge *et al.*, 1994). Partly, modern ranges postdate the Laramide (Chapter 8). Thus, the modern southern Rocky Mountains are a composite of uplifts formed over a span of some 60–70 Myr or so.

Jurassic to early Cenozoic compressional deformation probably caused considerable crustal thickening, with crustal thickness estimated to be 70 km or greater in the region immediately east of the Sevier orogen by the end of the Laramide. Shortening of the *upper* crust was accomplished by thrusting, whereas shortening of the *middle* and *lower* crust probably occurred by ductile strain. Ductility of the crust was facilitated, especially in the western part of the Cordillera, by magmatism associated with the arc, which had the effect of heating and softening the crust (Burchfiel *et al.*, 1992). As discussed in the next chapter, crustal thickening played a critical role in subsequent middle to later Tertiary extensional deformation.

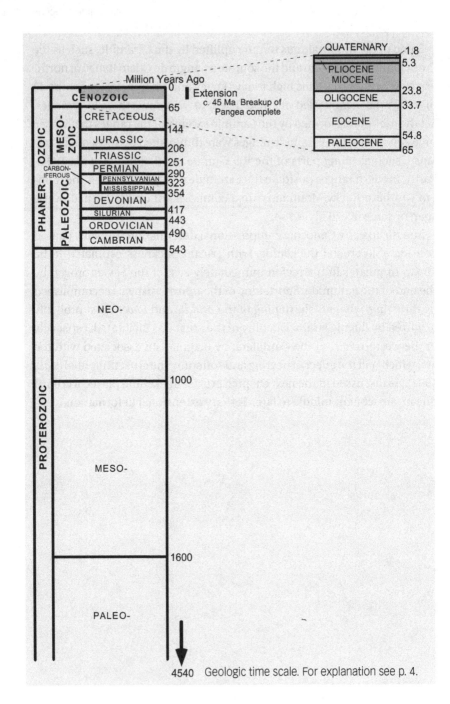

QUATERNARY — 1.8
PLIOCENE — 5.3
MIOCENE
OLIGOCENE — 23.8
EOCENE — 33.7
PALEOCENE — 54.8
65

Million Years Ago

Extension
c. 45 Ma Breakup of
Pangea complete

PHANEROZOIC			
	CENOZOIC		0
	MESOZOIC	CRETACEOUS	65
		JURASSIC	144
		TRIASSIC	206
	PALEOZOIC	PERMIAN	251
	CARBON-IFEROUS	PENNSYLVANIAN	290
		MISSISSIPPIAN	323
		DEVONIAN	354
		SILURIAN	417
		ORDOVICIAN	443
		CAMBRIAN	490
			543

PROTEROZOIC
NEO-
1000
MESO-
1600
PALEO-
4540

Geologic time scale. For explanation see p. 4.

The modern Southwest

Oligocene to present

Oddly shaped columns and spires emerge as erosion widens and removes rock along
joints in a formerly continuous sheet of ash-flow tuff. The tuff, exposed in the
Chiricahua Mountains of Arizona, was erupted from the Turkey Creek caldera
approximately 26.8 Ma (du Bray and Pallister, 1999; McIntosh and Bryan, 2000).

8.1 Introduction

The latest chapter in the geological development of the Southwest encompasses the most recent 35 Myr or so of the Earth's history. For perspective, recall that 35 Myr is less than one percent of the age of the Earth. Despite this short period of time, it is the time about which geologists know the most. The present chapter is defined principally by events following cessation of widespread crustal shortening in the western Cordillera. Extension and vertical crustal adjustments replaced shortening.

Despite the fact that the Earth's tectonic plates were nearly in their present relative positions in the Oligocene (Fig. 8.1), a profound change in plate geometry occurred during the latest 35 Myr along the western margin of North America. The Farallon plate, whose subduction beneath the North American plate for more than 140 Myr created the series of nearly continuous orogenies in western North America, was gradually and nearly completely consumed. Only fragments of the original Farallon plate, all carrying different names, remain along the margins of North and South America. Disappearance of the Farallon plate had major implications for the tectonics in the interior of the plate, dramatically affecting the geology and landscape of the Southwest. The change in plate geometry and rela-

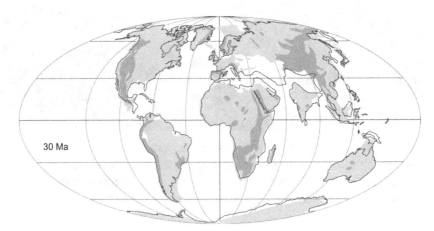

Fig. 8.1. Paleogeographic map showing the relative locations of the continents during the Oligocene epoch, approximately 30 Ma. Stippled pattern indicates land areas, with darker pattern indicating higher elevation. Modified from Smith *et al.* (1994).

tive plate motions resulted in massive magmatism and a different kind of orogeny, an 'extensional orogeny,' which radically reshaped the landscape. It is within this period of time that most of the modern landscape – the magnificent Rocky Mountains, the high and colorful Colorado Plateau country, and the dramatic and stark deserts of the Basin and Range province – formed, even though most of the rocks themselves are much older.

Throughout the Cenozoic, North America remained high relative to sea level, higher than at most other times during the Phanerozoic. Sedimentation in the Southwest was exclusively terrestrial, driven by the topographic relief formed by the Sevier and Laramide orogenies and by later lithospheric extension.

8.2 Death of a plate

The change in plate geometry began to occur at the boundary between the Farallon and North American plates about 28 Ma. At that time the trailing edge of the Farallon plate entered the subduction zone, bringing the Pacific plate into contact with the North American plate for the first time and signaling the beginning of the end of the Farallon plate. First contact occurred at about the latitude of the present USA–Mexico border. In contrast with the relative motion between the Farallon and North American plates, which was nearly perpendicular convergence, relative motion between the Pacific and North American plates was nearly parallel to the boundary, leading to right-lateral shear. These relations are difficult to visualize without a globe. Between the North American and Pacific plates, a transform boundary formed, at each end of which a 'triple junction' marked the point of contact of the Pacific, North American, and former Farallon plates. The Farallon plate by this time had broken into several smaller plates. The northern and southern triple junctions, respectively, are known as the Mendocino and Rivera triple junctions. By about 20 Ma, the Mendocino triple junction was located at about the latitude of San Diego (Fig. 8.2). As the Farallon plate slowly slipped beneath the North American plate, the transform zone progressively lengthened. Today the Mendocino triple junction, located at the common point between the Pacific, North American, and modern Juan de Fuca plates, coincides with the Mendocino fracture zone located offshore of northern California. The Rivera triple junction marks the common point

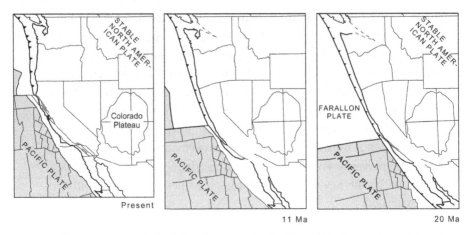

Fig. 8.2. Geometrical relations between the Pacific and North American plates 20 and 11 Ma, compared to present plate configuration. The relative extension and translation that occurred since 20 Ma is shown by the state outlines. Hachured line is the subduction zone. Modified from Atwater and Stock (1998).

of the Pacific, North American, and modern Cocos plates and coincides with the Rivera fracture zone just south of Baja California. Both the Juan de Fuca and Cocos plates are remnants of the Farallon plate. The notorious San Andreas fault of California is a segment of the lengthening transform boundary.

Exactly *how* the slab was subsumed is uncertain. Although subduction during the Cretaceous-to-early-Tertiary Laramide orogeny most likely occurred at a shallow angle, the subduction-angle almost certainly steepened during the Eocene, correlating with an episode of extension and arc-related magmatism (described below). Thus, one idea is that the slab simply resumed its slide into the mantle at a relatively steep angle, as characterized much of the previous 140 Myr-history of the Farallon plate, until it was fully consumed. Or, possibly the slab broke and imbricated slabs formed. If so, then both the old and new slabs simply slid into the mantle. Another, and innovative, possibility is that the slab folded along an east–northeast-trending axis (Fig. 8.3) before finally breaking and subsiding into the mantle (Humphreys, 1995). In any case, the effect of losing the Farallon plate from shallow-mantle depths caused dramatic changes in the dynamics of the Southwest. First, gradual loss of the plate removed the source of fluids and other 'incompatible' constituents supplied to the

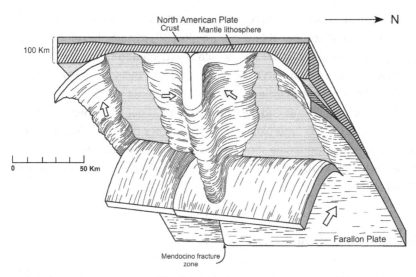

Fig. 8.3. Possible model of a buckled Farallon slab descending beneath the western USA about 35 Ma. This unusual view is looking *upward* toward the west from beneath eastern Canada. The front of the figure is approximately the longitude of Denver, Colorado. In this view the buckled slab has broken, and steer subduction is being re-established behind it along most of the western margin of North America. Direction and relative rates of motion of the slab with respect to North American are shown by arrows. Modified from Humphreys (1995).

overlying lithosphere, from which the arc-related melts were generated. Second, asthenosphere welled up to replace the plate. As it rose, it underwent decompression melting. In addition, because the asthenosphere was several hundred degrees hotter than the subducting oceanic lithosphere it replaced, it was capable of partly melting the overlying lithosphere. Both processes led to the present, mostly basaltic compositions of melts.

So, where is the Farallon plate now? As the plate slipped beneath the North American plate, it disappeared from direct observation. Its existence can only be inferred from the remaining remnants, the Juan de Fuca and Cocos plates; from patterns of magmatic anomalies on the Pacific sea floor; and from spatial and temporal patterns of magmatism and tectonic deformation in western North America. Until a few years ago, it would have been impossible to image the subducted plate, but recent advances in seismic tomography have made it possible to interpret the existence of the plate all the way to the base of the mantle. The presence of the plate is inferred from seismic velocity anomalies. 'Fast' seismic velocities are interpreted to represent the slab, which is expected to be cooler than the adjacent mantle. Fragments of the trailing edge of the plate, still 200° or so cooler than ambient mantle temperatures, underlie the Southwest from Arizona to eastern Texas and the Gulf of Mexico at depths between 350 and 650 km (van der Lee and Nolet, 1997). At greater depths, seismic P- and S-wave velocity anomalies allow the plate to be traced to the core-mantle boundary, 2700 km deep, beneath the western Atlantic Ocean (Grand *et al.*, 1997).

8.3 Volcanic outburst

Following the Sevier–Laramide orogeny, a regional pattern of erosion and sedimentation developed, as areas of previously high elevation began to wear down. But this stability was soon disrupted by an intense period of intermediate-to-silicic magmatism and by widespread faulting. The middle Tertiary was a time of widespread and voluminous magmatism over a large region of northwestern Mexico and the southwestern USA. In general, volcanism is older in the northernmost and southernmost Basin and Range province, becoming younger in the central Basin and Range (Fig. 8.4). Broadly, magmatism was either slightly older than or correlative with widespread extension, described in the next section. However, in many areas magmatism occurred during times of reduced extensional tectonism (Sonder and Jones, 1999), and details of timing and extension differ in different areas. An extensive belt of large, Upper Eocene-to-Oligocene silicic volcanic fields was formed at this time, stretching northward almost without break from central Mexico to central New Mexico (Fig. 8.5). The major volcanic field of this belt forms the Sierra Madre Occidental of northwestern Mexico, possibly the most extensive silicic volcanic field in the world (Keller *et al.*, 1982). It covers an area of approximately 296,000 km^2 of western Mexico (Swanson and McDowell, 1984), an area slightly greater than the State of Arizona. The Sierra Madre is a partly dissected volcanic plateau 1200 km long, which parallels the continental margin. The axial portion of the range is 200–300 km wide, although flanking

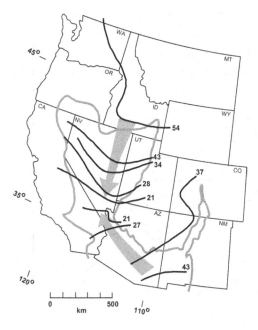

Fig. 8.4. Propagating post-Laramide magmatic fronts. Heavy lines show advance of magmatism toward southern Nevada from the north across Nevada and from the southeast across New Mexico and Arizona at the times indicated (in Myr). The gray line indicates the current boundary of the Basin and Range province and the Rio Grande rift, both regions that underwent significant late Tertiary crustal extension. From Humphreys (1995).

magmatic rocks cover a much wider region. Magmatism in the Sierra Madre was nearly continuous from Laramide through middle Tertiary time. The lower part of the sequence (Cretaceous–Eocene) consists mainly of andesitic flows; the upper part is dominated by Oligocene rhyodacitic ash-flow tuffs erupted from numerous large calderas (Clark *et al.*, 1982; Keller *et al.*, 1982; Luhr *et al.*, 2001). Although only a dozen or so calderas and caldera complexes are recognized, altogether several hundred may be present at different levels in the volcanic section (Swanson and McDowell, 1984). The detailed work required to elucidate these calderas has not been done.

This volcanic belt continues into southeastern Arizona and the 'Boot Heel' region of southwestern New Mexico. The Boot Heel volcanic field (Fig. 8.6) consists of at least nine large, deeply eroded calderas, ranging in age from 35.2 to 26.8 Myr, erupted in two separate pulses (McIntosh and Bryan, 2000). The calderas represent catastrophic eruptions, producing voluminous rhyolitic ignimbrites, which ponded within the calderas and also flowed outward, forming regional outflow sheets (Fig. 8.7). The volumes of these large ash-flow tuff sheets range from as small as 35 km^3 to as large

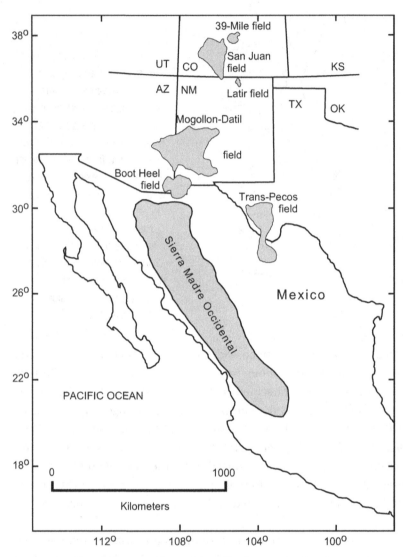

Fig. 8.5. Map of major Late Oligocene to Eocene silicic volcanic fields in western North America. From Clark *et al.* (1982) and McIntosh and Bryan (2000).

as 650 km³. In addition to rhyolitic ignimbrites, the volcanic field includes basaltic-andesite to rhyolitic lava flows (McIntosh and Bryan, 2000). The Boot Heel field is a major volcanic field, covering an area of more than 24,000 km². Yet, it has been very difficult to study, in part because the area is now part of the Basin and Range province. Parts of the calderas, and

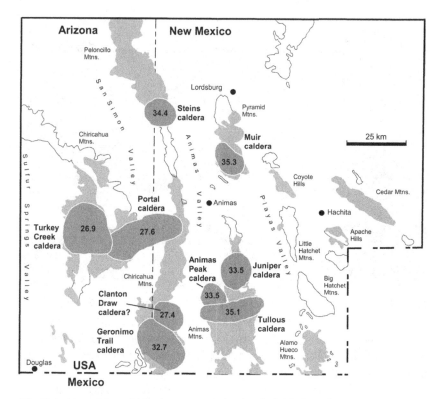

Fig. 8.6. Map of the mid-Tertiary Boot Heel volcanic field of southwestern New Mexico. Medium gray shading is deeply eroded calderas, with ages indicated in Myr. Light gray shading indicates approximate extent of volcanic outcrops. The Boot Heel volcanic field is heavily disrupted by basin-and-range faulting, which makes reconstruction of calderas difficult. Modified from McIntosh and Bryan (2000).

especially of their extensive ignimbrite outflow sheets, have been buried beneath middle and upper Cenozoic sediments, thus disrupting and obscuring the exposure and continuity of magmatic units. In places, uplift along Basin and Range faults has promoted deep erosion and exposure of calderas and tuff units (see photo at beginning of this chapter). Understanding of this field has also been hindered because, until recently, precise radiometric dating methods were not available to correlate distant outcrops of tuffs with their respective sources (McIntosh and Bryan, 2000).

Fig. 8.7. The Oak Creek Tuff is an outflow ignimbrite erupted 33.5 Ma from the Juniper caldera in the Animas Mountains. The Oak Creek Tuff is characterized by abundant phenocrysts of quartz, sanidine, plagioclase, and hornblende. Flattened pumice clasts (fiamme) and gas cavities (lithophysae) are apparent in this photograph. The fiamme are whitish in color compared with the pinkish color of the crystal-rich, glassy matrix, and are more vesicular and/or vuggy compared with the matrix. The scale is oriented parallel to the flattening direction of the fiamme. This outcrop is located near the northern Pyramid Mountains, approximately 58 km north of the source of the Tuff (Fig. 8.6) (McLemore *et al.*, 2000).

The Boot Heel volcanic field is probably contiguous with the Mogollon–Datil volcanic field (Fig. 8.5), although now separated by faulted, sediment-filled basins. The Mogollon–Datil field lies in the transition zone between the Colorado Plateau and the Basin and Range province. It comprises the rugged and beautiful Gila wilderness, the central part of which is a volcanic plateau 125 km in diameter with a mountainous rim and an interior basin. Great masses of flow-banded rhyolite characterize much of the rim. Several of the older calderas are preserved only in walls of younger calderas, which cut and offset earlier calderas. Magmatism ranged from 34 Myr old, about the time that early extension occurred in the central New Mexico area, to about 22 Myr old. The calderas here, as well as in the Boot Heel volcanic field, overlie shallow plutons in the upper crust, which together make up a composite granitic batholith.

From the Mogollon–Datil volcanic field, the extensive belt of mid-Tertiary silicic volcanic fields continues northward into northern New Mexico and southern Colorado (Fig. 8.5). In Colorado, the San Juan volcanic field, consisting of 15–20 major ash flow eruptions and probably at least 15 deeply eroded calderas, is a counterpart to the Mogollon–Datil volcanic field, but because of its mineral wealth has been much better studied (Lipman, 1992). Eastward from the Sierra Madre Occidental, middle Tertiary volcanic rocks extend across the Mexican Basin-and-Range province into west Texas, where they comprise the Trans-Pecos volcanic field. The main phase of volcanism in Trans-Pecos occurred 38–32 Ma. Volcanism was volumetrically dominated by at least 11 known or suspected calderas, some as small as 3–4 km in diameter, that erupted rhyolitic ash-flow tuffs. Typically, calderas were filled by lavas of intermediate to mafic compositions. In addition, numerous lava flows and small intrusions were emplaced contemporaneously with but independently of the calderas (Price *et al.*, 1986; Henry and Price, 1986).

The compositions of the widespread and voluminous intermediate to silicic magmatic rocks are calc-alkaline in their characteristics, hence are related to the subducting Farallon slab. They represented, then, a very broad continental magmatic arc. Isotopic compositions of some of the most silicic magmatic rocks, including the widespread ash-flow tuffs, are compatible with their derivation, in part at least, from remelted Proterozoic crust, indicating that the lithosphere was very hot at this time. Near the eastern edge of the broad region of middle Tertiary magmatism, such as the Trans-Pecos volcanic field of west Texas, compositions are significantly more alkaline. In this field, a strong compositional gradient is present. In the western part, compositions were calc-alkalic, generally similar to those of the Sierra Madre Occidental, Boot Heel, and Mogollon–Datil volcanic fields. East and northeastward in the volcanic field, farther from the plate margin where the subducting plate was deeper and the overlying mantle wedge drier, compositions became strongly alkalic, in contrast with most other volcanic rocks of this time interval (Barker, 1979). The slab-driven volcanism of the Sierra Madre Occidental and related volcanic fields continues today in the east–west-trending Mexican volcanic belt of central Mexico (Nieto-Samaniego *et al.*, 1999; Ferrari *et al.*, 1999).

8.4 A new regime

The crustal deformation that generally accompanied the middle Tertiary magmatism described in the previous section was characterized, not by crustal shortening as for most of the previous 250 Myr, but by extension, as the continent was pulled apart in a new stress field. Typically, extensional deformation in the Southwest is described in terms of a two-stage history, with an earlier (Oligocene) phase of low-angle faulting and a later (Late Miocene–Pliocene) phase of mainly high-angle normal faults. Initially, crustal extension occurred in broadly linear zones, belts no more than a few hundred kilometers wide corresponding generally with regions of overthickened crust behind the Laramide fold and thrust belts. Initiation of extension in the Cordillera was diachronous. Although generally older in the northern Cordillera (46–45 Myr), extension in the Southwest began in the earliest Oligocene, around 36 Ma. Unlike in the northern Cordillera, where extension occurred within or outboard (west) of the tectonically thickened Paleozoic continental margin, extension in the Southwest was superimposed on thickened cratonic crust (Christiansen and Yeats, 1992; Sonder and Jones, 1999). Extension throughout the Southwest was typically preceded and accompanied by intermediate to silicic volcanism (next section).

One such narrow extensional region stretches from the Colorado River in southeastern California southward and eastward through southern Arizona, ending in northern Sonora (Fig. 8.8). Among the expressions of this extension are terranes denoted as 'metamorphic core complexes.' Core complexes are characterized by exposure of metamorphic and plutonic rocks (the 'core'), typically with a gneissic or mylonitic fabric, separated from younger, non-metamorphosed stratified cover rocks by low-angle shear zones, or detachment surfaces (Figs. 8.9, 8.10). The cover rocks are typically cut by closely spaced normal faults, commonly with low dips. Fault blocks have usually undergone rotation. The high-grade metamorphic and plutonic rocks underlying the detachment surfaces are thought to represent middle crust, 10–15 km deep, that has been exposed by large relative extension and crustal thinning (Christiansen and Yeats, 1992). Considerable difference of opinion exists among geologists regarding the amount of lateral displacement on the low-angle detachment surfaces and on their role in exhuming the middle crust (Christiansen and Yeats, 1992).

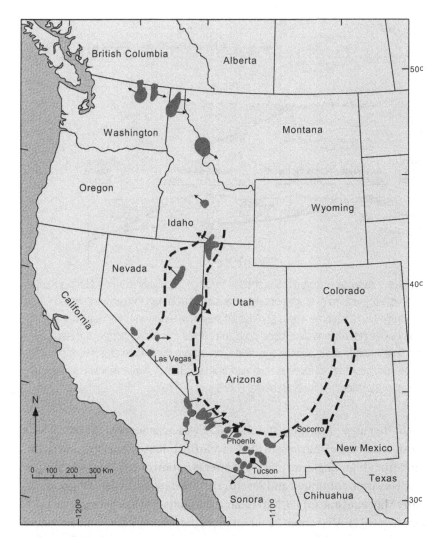

Fig. 8.8. Extensional zones of Oligocene and Miocene age. Dashed lines delimit extensional belts of the eastern Great Basin, southern Basin and Range province, and Rio Grande rift. Dark gray color indicates metamorphic core complexes. Arrows show trend and vergence of kinematic indicators in selected core complexes. Modified from Christiansen and Yeats (1992).

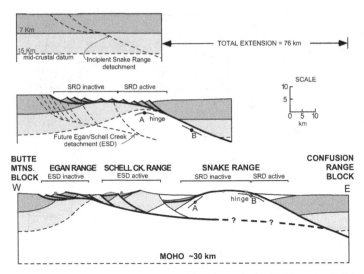

Fig. 8.9. Stepwise-restored cross section through the highly extended Snake Range domain. Progression of stages during extension is from top to bottom. Different low-angle faults became active at different stages of extension. Although not all highly extended regions developed core complexes, the Snake Range domain did (the Snake Range itself), where crystalline rocks from the middle crust were brought to the surface. Hinge in footwall migrated from position A in intermediate stage to B in final stage. Modified from Wernicke (1992).

Another area of early crustal extension was near Socorro in central New Mexico, the area of the present Rio Grande rift. Initial extension occurred on numerous high-angle normal faults after eruption of a 31 Myr-old ash-flow tuff. However, continued and severe crustal extension in Late Oligocene to Early Miocene time (31–20 Ma) rotated the faults into a subhorizontal position (Fig. 8.11). After about 30° of rotation, the original faults were no longer favorably oriented for continued slip, hence a second set of Middle to Late Miocene high-angle normal faults was initiated. The earliest alluvial sediments associated with the extension are 27 Myr old. Successively younger strata have respectively lower dips (Fig. 8.11). This *rotation* of initially high-angle faults to a subhorizontal position in the Socorro area is in contrast with many of the low-angle faults associated with core complexes, which were apparently *active* at a low angle. The total strain over the past 30 Myr in this local area amounts to approximately 200% (Chamberlin, 1983), compared with approximately 50% in the Socorro region more generally

Fig. 8.10. A detachment fault in the Newberry Mountains near Laughlin, Nevada, is well exposed, as shown here. Allochthonous upper plate rocks consist of coarse-grained granite and volcanic rocks, weathered to a reddish color. Rocks of upper plate have many small faults and are shattered. They are separated from autochthonous fine-grained granite in the lower plate by a knife-sharp, subhorizontal detachment fault. Bottom of scale rests on thin (5–10 cm thick), whitish microbreccia unit underlying the fault surface. At this location upper plate rocks were transported approximately 24 km to the northeast. The age of faulting is inferred to be about 12 m.y. Reference: Mathis (1982). Scale is 51 cm.

(Chapin and Cather, 1994). Despite the large amount of local extension, no core complex was formed here.

From the areas of initial high strain, extensional deformation became widespread throughout the Southwest (but not on the Colorado Plateau) after the Middle Miocene, continuing to the present. This later phase, described as the basin-and-range phase of extension, gave rise to the present physiography of the Southwest (Christiansen and Yates, 1992) (see Section 8.6). Extension did not significantly affect the region of the Colorado Plateau, presumably because the lithosphere was too cold and strong. In contrast with the earlier, more classical idea that the western USA attained its present high elevation only recently (latest 10–15 Myr), it is now thought likely that in the early and middle Tertiary the western USA resembled a wide, high mountain belt, similar to the modern Chilean Andes. During the basin-and-range phase of extension, the western USA finally

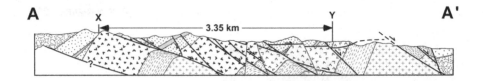

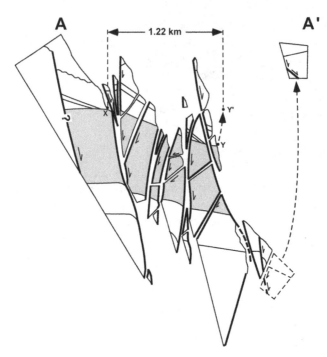

Fig. 8.11. Upper panel: Simplified geologic cross section (no vertical exaggeration) through the Lemitar Mountains near Socorro, New Mexico. The geometry of low-angle faults with respect to pre-rift stratigraphic units indicates that the faults were initiated at a high angle and rotated to a subhorizontal attitude in the Late Oligocene to Early Miocene. A second generation of high-angle faults was initiated in the Middle to Late Miocene. The overall extension compared to the pre-extension geometry (lower panel) is 275%. Lower panel: The same area as above restored to original pre-extensional geometry. Shaded area corresponds to 'v' pattern (volcanic rocks) in upper panel. The restoration was made by systematically unfaulting and untilting the superimposed fault blocks of the present-day section. The points X and Y are the same reference points as in the unrestored section (upper panel). The apparent reversal of dips on normal faults (apparent compression), results from downwarping of the section to the east, with resultant normal drag and rotation between the closely space fault blocks. It is removed by projecting Y to the horizon X–Y' along an arc radiating from the hinge at X. Modified from Chamberlin (1983).

collapsed, reaching its present, and *lower*, elevation. As recently as 16–15 Ma, western Nevada stood approximately 3 km above sea level (higher than Quito, Ecuador, at about 2.9 km), but by 13–12 Ma it had attained its present elevation of 1–1.5 km. Even the southern Rocky Mountains in southwestern and central Colorado were much (1–1.5 km) higher than at present (Wolfe *et al.*, 1997, 1998). Probably the Colorado Plateau, which by 21–20 Ma was not yet structurally differentiated from the Basin and Range province, attained its structural and physiographic definition (but remained high) during this Miocene collapse event. This widespread phase of extension was accompanied by basaltic volcanism, which, since the beginning of the Quaternary, has tended to focus near the margins of the extended regions. The axis of extension, which had been oriented approximately northeast–southwest during early extension (Oligocene) also rotated to become more northwest–southeast during later phases of extension (Middle Miocene–Pliocene).

8.5 Legacy of the Laramide

What caused this widespread extensional deformation and the outburst of magmatism that accompanied it, both of which clearly overlapped with subduction of the Farallon plate? At the present time, no general agreement exists among geologists regarding the cause or causes, but several explanations have been proposed. In the classical model, the control for tectonism and magmatism derives from relative plate motions among the North American, Pacific, and Farallon plates. Lithospheric strain resulted from a combination of boundary forces acting on the margins of the plates and forces acting on the bases of the plates. This model recognizes that the rate of closure between the Farallon and North American plates decreased about mid-Tertiary time approximately 40 Ma (Sonder and Jones, 1999), suggesting that the dip angle of the subducting slab suddenly steepened. Because the downgoing slab was again (after the late Laramide–Sevier phase of flat-slab subduction) deep enough in the mantle for melting to occur, arc magmatism resumed as the sinking slab carried water and other volatile components deep enough into the mantle to flux the overlying (and newly introduced) asthenospheric wedge.

Reestablishment of the downgoing Farallon plate closer to the North American plate margin required counterflow of asthenospheric mantle into the region above the plate, heating, weakening, and buoying the base of the continental lithosphere and driving extension and magmatism that gradually re-established arc volcanism (Lipman, 1992). An earlier and now less favored variation of this model suggested that, rather than gradually increasing its dip angle, the sinking slab may have broken apart and reestablished itself farther to the west at a steeper angle (Lipman, 1980). Another variation of the model does not require a changing slab geometry nor imbricated slabs, but simply postulates that the mid-Tertiary magmatic event reflected back-arc extension (possibly due to decreased convergence or rollback of the slab) (Elston, 1984; Lipman, 1980). Evidence for these models stems from the temporal pattern of magmatism in the southern Basin and Range province, where renewed magmatism was oldest toward the east, in New Mexico, southern Colorado, and west Texas, and swept progressively westward. However, none of these models addresses the fact that volcanism is older in both the northern and southern Basin and Range province than in the central segment (Fig. 8.4). Steepening of the slab by downward buckling or double-sided rollback along an east–northeast-trending axis could explain the observed trends in younging of volcanism toward southern Nevada (Fig. 8.3) (Humphreys, 1995). Other explanations offered to explain magmatism with associated extension included detachment of overthickened mantle lithosphere with consequent replacement by asthenosphere (similar to above), and impingement of a plume on the base of the lithosphere. In addition, after 28 Ma a slab 'window' may have opened behind the Farallon plate, a consequence of cessation of sea-floor spreading as the East Pacific rise intersected the trench along the western margin of North America. The 'footprint' of the gap on the base of the overlying lithosphere is a triangle that increases in size with time. Upwelling of asthenosphere into the widening hole potentially provides heat that weakens and buoys the overlying lithosphere and provides a source of magma that accompanies associated extension. However, because the Farallon plate was young and warm at the time of subduction, it was probably nearly indistinguishable thermally from the asthenosphere. Other problems with size and timing of the slab window exist (Sonder and Jones, 1999).

In an alternative model, the Middle Eocene and younger extension and magmatism may be related to stresses arising from *within* the North American plate itself (Wernicke *et al.*, 1987; Livaccari, 1991; Sonder and Jones, 1999). Sevier and Laramide crustal shortening caused crustal thickening and consequent elevation increase, which affected the continental crust as far east as the Sangre de Cristo and Front Ranges of New Mexico and Colorado, which are now up to 1500 km from the plate margin. Not only was the upper crust shortened and upper-crustal rocks pushed in great thin-skinned thrust sheets and basement-cored uplifts over underlying and younger rocks, but also the lower crust was thickened. The crust may have been up to 70 km thick in parts of the Cordillera (McQuarrie and Chase, 2000). Although overthickened crust is initially relatively cold and therefore strong, it gradually loses its strength as heat is conducted into it. The over-thickened crust could not be sustained. Despite the fact that subduction along the western plate margin, ceasing progressively from south to north, continued throughout most of the Tertiary, simultaneously the Laramide orogen began to collapse, spreading outward and downward under its own weight in response to gravitational forces. This mechanism is described as 'gravitational collapse' or 'gravitational thinning' of the lithosphere. Collapse occurred first in regions where crustal thickening coincided most closely with Cretaceous batholiths, such as in the Pacific Northwest and northern Rocky Mountains. In these areas the lithosphere was initially hot and therefore weak. In other areas, such as the northern and southern Basin and Range province, where the crust was only sparsely intruded, the delay was longer because more time was need for heat to infuse into and weaken the lithosphere. Areas where thickening was accompanied by little plutonism, such as the central Basin and Range province, required the longest time to weaken sufficiently. In the Basin and Range, the heat source may have been provided by upwelling of the asthenosphere in the lee of the subducted Farallon plate (next section) (Wernicke, 1992). Thus, gravitational collapse is an inevitable result of crustal thickening. Ironically, then, the legacy of the Sevier–Laramide event, which caused major and far-reaching shortening throughout the West, was formation of an *extensional* stress field: a reversal of the original stress field which created the uplifts. The process of extension and magmatism continues throughout much of the Southwest.

8.6 The modern Southwest

Thus, we come to the *modern* Southwest, which comprises the tectonic provinces shown in Fig. 8.12. These provinces acquired their physiography during the widespread Late Miocene to Holocene basin-and-range phase of extension. The eastern part of the region, more or less coincident with the High Plains of New Mexico, Colorado, and much of Texas, comprises the Great Plains tectonic province. Characterized by a crust 50 km thick and a lithosphere in excess of 150 km thick, it is part of the stable craton of North America, which, except for deformation during the Ancestral Rocky Mountains orogeny (Chapter 5), has been relatively undisturbed since crustal formation (Chapter 1). The Colorado Plateau is a piece of the craton that has been rifted away from the Great Plains as the Rio Grande

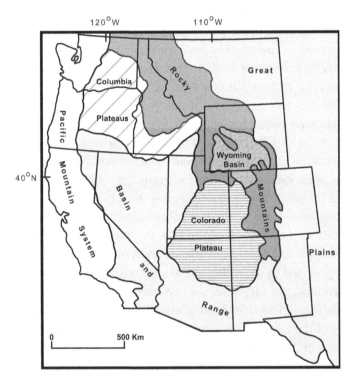

Fig. 8.12. Major physiographic and tectonophysical provinces of the Southwest, as commonly defined. Different investigators have delineated physiographic provinces differently. This map is simplified from Fenneman (1928). After figure in Eaton (1979).

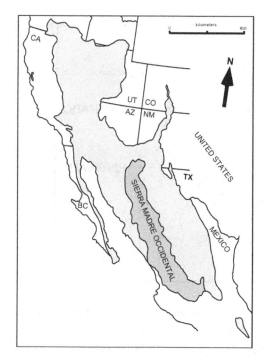

Fig. 8.13. The region of the Southwest that underwent significant middle to late Cenozoic lithospheric extension. The area includes the Basin and Range and Rio Grande rift tectonic provinces (Fig. 8.12). Although broken by extensional faults, the Sierra Madre Occidental range is a relatively unextended block separating areas of major extension in northern and central Mexico from those adjacent to the Gulf of California. From Eaton (1979) and Henry and Aranda-Gomez (1992).

rift has opened. It is a block of crust that remained relatively undeformed during the Mesozoic orogenies. Its best known physiographic feature is the Grand Canyon of the Colorado River.

The Southwest as we usually think of it occupies only the southern part of the Colorado Plateau. South and west of the Colorado Plateau and the Great Plains, the continental lithosphere has been stretched and thinned, resulting in a distinctive physiography of narrow mountain ranges separated by broad, sediment-filled basins. This region is the Basin and Range province, encompassing an area of greater than 1.8 million km² of western North America (Henry and Aranda-Gomez, 1992). In breadth it stretches from west Texas westward across New Mexico and Arizona to Nevada and California. Its physiography typifies much of the Southwest. In length, the Basin and Range province stretches 3000 km from Idaho and Oregon, far beyond the Southwest, southward across western and central Mexico nearly to Mexico City (Fig. 8.13). Extension may originally have reached as far south as the State of Oaxaca. The Mexican segment of the Basin and Range province is bisected by the Sierra Madre

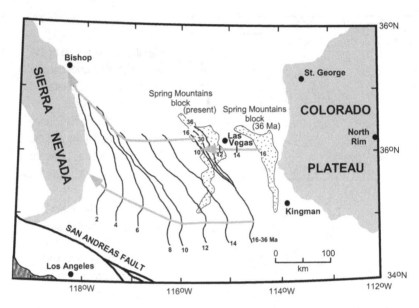

Fig. 8.14. Tertiary extension of the central Basin and Range province is shown by successive positions of the Sierra Nevada (indicated by solid lines) and the Spring Mountains (Nevada) relative to the Colorado Plateau since 36 Ma. Arrows indicate paths of selected locations. A major decrease in the elevation of the Basin and Range province occurred between 16 and 13 Ma (see text). Numbers are ages in Myr. From Wernicke and Snow (1998).

Occidental (Henry and Aranda-Gomez, 1992). In the USA, the Basin and Range province includes the region mainly of Utah and Nevada known as the Great Basin, which is defined by its interior drainage. The Basin and Range province, although relatively high topographically, has a thinner crust and lithosphere and hence a lower average elevation than the Colorado Plateau and Great Plains. The total amount of extension across the Basin and Range province, including all extensional deformation from the middle Tertiary to the present, is 250–300 km, mainly since 16 Ma, with an average extension direction of N73°W. That is, the Sierra Nevada has moved 250–300 km west-northwestward relative to the Colorado Plateau (Fig. 8.14). Considering that the Basin and Range province is about 350 km wide at the latitude of Las Vegas (Nevada), this amount of extension represents 400–500% (Wernicke and Snow, 1998). Most of the present-day deformation of the Basin and Range province occurs near the western and eastern boundaries (Thatcher *et al.*, 1999). Thus, the present elevation difference between the Basin and Range and the Plateau was probably created by spreading, lithospheric thinning, and topographic *collapse* of the Basin and Range, creating the impression that the Plateau rose in elevation.

Cenozoic extension of the western USA caused the Colorado Plateau to pull away slightly from the craton, leaving a linear series of faulted basins, thinned crust, and thinned lithosphere stretching from central Colorado to central New Mexico between: the Rio Grande rift (Baldridge *et al.*, 1995). From central Colorado to southern New Mexico the Rio Grande rift occupies the axis of a broad regional uplift, 2500 km in breadth and 1300 km in length. The uplift, which includes the Colorado Plateau and the western Great Plains, is similar to a mid-oceanic ridge in physiography and scale. It is most likely upheld by warm, buoyant asthenosphere (Lerner-Lam *et al.*, 1998). The crestal region of this enormous uplift in places exceeds 4.2 km in elevation. It defines the *tectonic*, in contrast with *physiographic*, southern Rocky Mountains (Eaton, 1986). South of Socorro, the rift merges with the eastern Basin and Range province. Because of its control of drainage patterns, the Rio Grande rift has had a major effect on both prehistoric and modern settlement patterns in the Southwest.

With the Farallon plate effectively out of the picture, the tectonics of the modern Southwest (i.e. Late Miocene to present) is dominated by two major stress regimes. First is the right-lateral shear generated by grinding of the Pacific plate along the western margin of the North American plate. This stress field is propagated into the North American plate at least as far as the Death Valley and Las Vegas (Nevada) regions, where faulting typically has a right-lateral component. The second is collapse and extension of the Basin and Range and Rio Grande rift, which continues even today, although at a reduced rate compared to the Oligocene and Miocene. Although deformation throughout the West and Southwest has been more or less continuous from the early or middle Tertiary to the present, it has become widespread only after the Middle to Late Miocene (the basin-and-range event). In contrast with earlier deformation, faulting of the basin-and-range deformation was mainly high-angle.

Although no consensus exists among Earth scientists regarding the exact cause of extension, a major component probably results from gravitational collapse, as discussed earlier in this chapter with respect to the Laramide. Gravitational collapse results from differences in **gravitational potential**

energy (Box 8.1) between adjacent regions of lithosphere. Whether large contrasts in potential energy can drive extension depends upon crustal and mantle temperatures and compositions of the lithosphere, among

---

Box 8.1 Gravitational potential energy. The potential energy stored in a volume of rock by virtue of its position in the Earth's gravitational field. For a column of continental lithosphere, the gravitational potential energy is the force at the base of the column (equal to the product of the mass of the column and the gravitational acceleration) times its height above some reference surface. The **buoyancy** force acting between adjacent columns of lithosphere arises from horizontal differences in density and thickness, equaling the sum of the pressure (force per unit area) difference between the two columns. The buoyance force may be non-zero even if the two columns are in isostatic equilibrium, in which case the buoyancy takes the form of the difference in the gravitational potential energy per unit area between the two columns. Lithospheric buoyancy depends most strongly on crustal thickness, which may be the major driving force for extension in the Southwest. That is, the higher a volume of rock, the greater its potential energy. The effect is that a tall mountain range (such as the modern Himalaya) will deform and collapse under its own weight (a response to the Earth's gravitational field). However, the buoyancy of the lithosphere is also significantly affected by lateral differences in crustal and mantle density between the columns, such as by warm, low-density mantle lithosphere. Numerical treatment of buoyancy may be found in Sonder and Jones (1999).

---

other factors, which determine the strength of the lithosphere. Regional variations in potential energy in the Southwest correlate well with modern deformation regimes, generally predicting the styles and rates of observed deformation (Jones *et al.*, 1996; Sonder and Jones, 1999). From recent studies, geologists infer that **buoyancy** forces (Box 8.1) arising from differences in potential energy among different tectonic provinces are capable of producing most of the active deformation observed in the Southwest. Differences in potential energy are greatest in the actively extending Great Basin and Rio Grande rift and are relatively low in the southern Basin and Range and Colorado Plateau provinces. In the Basin and Range, a small contrast in potential energy correlates with the observed low strain rates. In the Colorado Plateau with its thicker crust and cooler temperatures, small differences in potential energy are too low to drive significant extension. Contrasts in potential energy arise from different sources in different areas. In the Basin and Range province and probably also the Rio Grande rift, buoyant, asthenospheric mantle is responsible for extensional deformation. In contrast, in the Rocky Mountains much of the potential energy is stored in the crust, probably emplaced during the Laramide. This model is not accepted by all geologists (Lipman, 1992), and discussion continues regarding the cause or causes of the profound middle Tertiary tectonic and magmatic events. Despite the strong support for the importance, even dominance, of buoyancy forces in driving extension in the western USA, its overall role with respect to other forces is not fully understood. It is certain that buoyancy is only one of several forces (including plate boundary forces, which are especially important in the Great Basin and along the California coast) affecting extension in the Southwest (Sonder and Jones, 1999; Thatcher *et al.*, 1999; Flesch *et al.*, 2000; Oldow *et al.*, 2001).

From recent work using both seismic and geochemical techniques, geologists infer that the lithosphere beneath the Basin and Range province and the central and southern Rio Grande rift is significantly thinner than beneath the craton of the adjacent Great Plains and Colorado Plateau. With the exception of a region of southern Nevada, basaltic rocks erupted within the past five million years in the Basin and Range and southern rift provinces were generated from a region of mantle that has undergone long-term depletion in elements that readily partition into melts. This region is interpreted as *asthenosphere*. Conversely, young basaltic rocks from cratonic regions were generated at about the same depth from mantle that has undergone long-term enrichment in these same elements, leading to the interpretation that they were partial melts of *lithosphere*. Thus, lithospheric mantle appears to be thinner beneath regions of the Southwest that have undergone significant crustal extension. In southern Nevada, an 'iceberg' of thick lithosphere rifted away from the Colorado Plateau, apparently with little thinning. An important implication is that lithospheric mantle beneath the Southwest was not removed during the Sevier–Laramide period of low-angle subduction (Perry *et al.*, 1987; Livaccari and Perry, 1993).

Thus, the Southwest is a region underlain by shallow, hot asthenosphere, probably still turbulent in the wake of the Farallon plate as remnants of the slab sink toward the core. It is a region whose average high elevation is in part dynamically supported by hot, buoyant, asthenospheric mantle and in part by thickened crust remaining from Laramide shortening. The consequence of both is to create extensional forces and to promote gravitational collapse. Thus, the *landscape* of the Southwest is *modern*, although many of the rocks are, as discussed in earlier chapters, *ancient*.

The consequences of such deep-seated instability are inevitable and far-reaching: faulting and volcanism. The Southwest, although not nearly as active as areas adjacent to the San Andreas fault system at the plate margin, is none the less subject to frequent and large earthquakes. Numerous earthquakes of magnitude 7 or greater have occurred in the past one hundred years. One of the most severe, but least appreciated, occurred on May 3, 1887, in the San Bernardino Valley of Sonora, Mexico, a mere 100 km south-southeast of Douglas, Arizona. The earthquake, which was felt from Santa Fe, New Mexico, to Mexico City (Fig. 8.15), ranks as one of the greatest historic events of the western USA exclusive of California (DuBois and Smith, 1980). Although only of (moment) magnitude 7.2, it caused numerous deaths and extensive damage to many villages in northern Sonora. Ground breakage occurred over a distance of 50 km, with an average observed vertical displacement on the fault scarp of 3 m. A repeat of this earthquake could have disastrous consequences for the now heavily populated areas of southern Arizona. Moreover, geologists recognize much evidence for other 'recent' faults, faults that offset the modern land surface. Possibly the current lull in seismicity throughout the Southwest does not reliably reflect the average seismicity of the past few hundred thousand years. How recent is 'recent?' Many studies are underway to decipher the seismic history of some of the faults. Geologists are probably many years away from reliable estimates of the recurrence history and, therefore, the probability of future activity on some of the more important fault zones. However, it does seem safe to predict that the hazard to human life and property will increase as the population of the Southwest increases.

Volcanism in the Southwest is also widespread and young (Fig. 8.16), and will recur. As in the past, volcanism is of two types: isolated, relatively quiet effusions of basaltic lavas, and explosive eruptions of more silicic compositions from polygenetic centers. The most recent eruption in the Southwest was that of 300-m-high Sunset Crater Volcano, part of the San Francisco volcanic field near Flagstaff, Arizona. Eruption, most likely witnessed by the local indigenous population, occurred in AD 1064–1065, blanketing the surrounding region with black basaltic ash. Numerous other examples occur. The McCartys flow near Grants, New Mexico, is about 3000 years

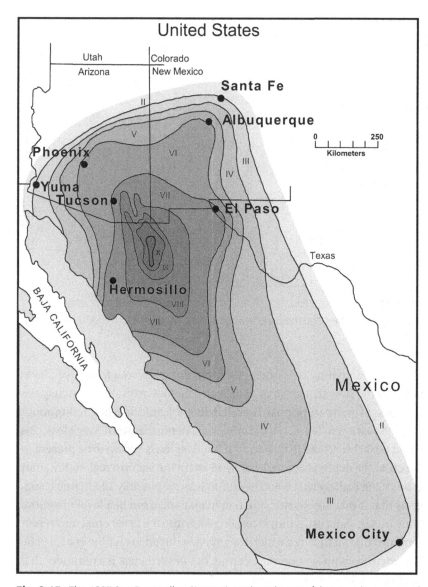

Fig. 8.15. The 1887 San Bernardino (Sonora) earthquake was felt over a large area of the Southwest, from Santa Fe, New Mexico, to Mexico City. The lines, called isoseismal lines, are contours of equal felt intensity. The exact locations of many of the isoseisms is uncertain, particularly in central Mexico. Roman numerals depict values of the Modified Mercali Intensity. From DuBois and Smith (1980).

Fig. 8.16. Map of late Cenozoic volcanic rocks in Arizona and New Mexico. All are younger than 15 Myr, and most are less than 5 Myr. Modified from Luedke and Smith (1978).

old, based on dating of burned plant remains recovered from a soil layer beneath the flow (Laughlin *et al.*, 1994). The McCartys flow is merely the youngest of many eruptions of basalt in this volcanic field beginning about 700,000 years ago (Fig. 8.17). Capulin Peak, in northeastern New Mexico, is about 60,000 years old (Sayre *et al.*, 1995). Magma is known to be present in the crust at a depth of approximately 19 km in the Socorro region (Rinehart *et al.*, 1979; Fialko *et al.*, 2001). Its sill-like form, possibly 100 m thick, suggests that it may be basaltic, the magma having risen to a level of neutral buoyancy in the crust. A root extending 6 km into the lower crust may represent the conduit through which magma was intruded (Schlue *et al.*, 1996). The Socorro chamber may never erupt, but nevertheless is a reminder that magmatism is active in the region.

Besides their generally basaltic compositions, these eruptions all have two features in common. First, the eruptions were localized and short-lived. That is, if the cinder cone or flow didn't directly destroy anything, there wasn't much more risk. Certainly, ash and cinders accompanied these eruptions, but significant deposits did not extend more than a few tens of kilometers from the vents. Eruptions probably did not last longer than a

Fig. 8.17. The Bandera flow, near Grants, New Mexico, is characterized by many partly collapsed lava tubes, as shown in this photo from the Big Tubes area of El Malpais National Monument. The Bandera flow, approximately 10,000 years old (Laughlin *et al.*, 1994), is representative of the widespread basaltic lava flows erupted widely in the Southwest since 5 Ma.

few months to a few years or tens of years. Second, the locations of impending eruptions probably cannot be predicted before magmas begin moving toward the surface. Although the likelihood is greatest that future basaltic eruptions will occur in the vicinity of previous eruptions, the fact that the Southwest is littered with the remains of 'one shot' eruptions – small eruptions that occurred far from any previous or subsequent outbreaks – suggests that similar eruptions could probably occur anywhere at any time.

Major silicic volcanism occurred in the geologically recent past in the Southwest and is most likely to recur. In the Jemez volcanic field of northern New Mexico (Fig. 8.18), major volumes of ash-flow tuffs were erupted 1.5 Ma and again 1.2 Ma (Fig. 8.19). They were accompanied by formation of calderas, the most recent of which, Valles caldera, is approximately 20 km in diameter. The combined volume of silicic tephra (recalculated to magma density) released during these two eruptions was 300–325 km^3 (G. Heiken, unpublished data), more than 30 times greater than the

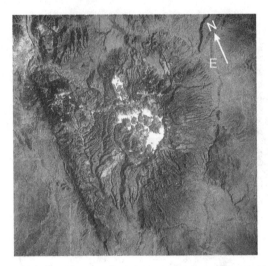

Fig. 8.18. The circular 1.2 Myr-old Valles caldera is clearly visible in this satellite image of the Jemez volcanic field of northern New Mexico. The topographic 'moat' is partly outlined by snow. The small domes within the moat, whose locations are interpreted to mark the main fracture along which the caldera subsided, were erupted sequentially between approximately 1.0 Myr and 50,000 years ago (Heiken *et al.*, 1990). The ring fracture is about 20 km in diameter. The domes consist of the same high-silica rhyolite as the ignimbrites, although are different in some geochemical properties. Near the center of the caldera is a resurgent dome, formed as pressure from the reactivated magma chamber pushed the floor of the caldera upward. The present topographic rim of the caldera is erosional. Enlargement of the caldera occurred, in part, while eruption was in progress as blocks of the footwall collapsed inward. The entire eruption during which the caldera was formed and approximately 325 km³ of tephra emplaced may have lasted only a few months or less (K. Wohletz, Los Alamos National Laboratory, personal communication). The linear mountain range west of the caldera is the Sierra Nacimiento, a fault-bounded uplift of Laramide age. E represents the city of Española. Width of image is approximately 95 km.

eruption of Krakatau in 1883 (18 km³ *not* reduced to dense-rock equivalent) (Simkin and Fiske, 1983) and 300 times greater than the 1984 blast of Mount St. Helens (< 1 km³). Judging by the amount and distribution of Valles ash spread across the USA, these eruptions certainly had regional and probably also global effects. Eruptions in the Jemez volcanic field continued intermittently until between 60 and 50 ka (Reneau *et al.*, 1996), although of

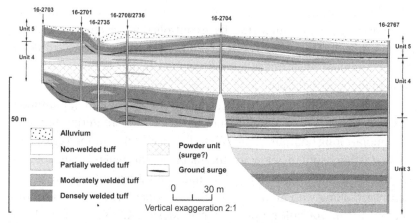

Fig. 8.19. The Bandelier Tuff comprises fallout and ash-flow tuff deposits that accompanied formation of the 20-km diameter Valles caldera in the Jemez volcanic field, New Mexico (Fig. 8.18) 1.2 Ma. In detail, the stratigraphy of the tuff is very complex. Mappable units are a combination of flow units, such as ground surge deposits, and 'cooling units.' Cooling units typically comprise packages of individual depositional units emplaced closely in time that cooled together. The degree of welding (sintering together of individual glass shards and pumice clasts) that characterizes cooling units is a function of the thickness of the flow, the cooling time before emplacement of next-overlying deposits, and of distance from the source. Porosity and permeability decrease, and density and devitrification increase, with increasing degree of welding. Erosional, weathering, and hydrological properties of tuffs such as the Bandelier are dominantly controlled by cooling units. Vertical lines indicate locations of core holes. The work illustrated here was conducted by K. Wohletz of the Los Alamos National Laboratory. Modified from a figure prepared by Wohletz in RCRA Facilities Investigation for Potential Release Site 16-021(c), Los Alamos National Laboratory unpublished report LA-UR-98-4101 (1998).

significantly diminished size. Travel-time delays of seismic waves from distant earthquakes recorded in and adjacent to Valles caldera convincingly demonstrate that a large volume of magma is still present in the middle crust at a depth of 5–16 km beneath the volcanic field (Steck *et al.*, 1998). Probably it represents refilling of magma chambers beneath the volcanic field. Whether it will result in future enormous and catastrophic eruptions, smaller and more localized eruptions, or any at all depends on numerous factors. The size and timing of future eruptions in the Jemez seem impossible to predict, although they will almost certainly occur.

In the foreseeable geological future, the Southwest will continue to be dominated by the same forces that have controlled the recent past. As parts of California, the USA, and all of Baja California slide northwestward along the San Andreas transform fault, the lithosphere in the Southwest will continue to undergo extension and thinning, although perhaps at a reduced rate. Basins will continue subsiding adjacent to high-standing mountain ranges, with occasional temblors such as that of the San Bernardino Valley, Sonora, along bounding fault zones. Because of the thinned crust and shallow asthenosphere below, geothermal gradients will be high. Volcanism will remain an important and dramatic, if infrequent, feature of the Southwest, along with attendant geothermal phenomena. Based on the fact that several volcanic eruptions in New Mexico, Arizona, and northern Mexico have occurred within the past several thousand years, more can be expected in the next few thousand.

Finally, the continuing growth of population in the Southwest is raising and will continue to drive a host of social issues related to the geology of the region. One of the most pressing current concerns is the availability and quality of groundwater, on which many major municipalities rely heavily or even exclusively. Future decisions regarding allocation of this precious resource will affect rural and municipal regions alike, as well as relations among states and with Mexico. Informed decisions will require much more detailed knowledge of the extent and properties of water-bearing aquifers than currently exists. Other issues, such as the disposal of hazardous and non-hazardous wastes, soil stability, flooding hazards, and extraction of hydrocarbon and mineral resources, rather obviously require geological knowledge. Much data already exist on which to base decisions, but as population grows and more resources are required – i.e. as these issues become more pressing – more information will be required. Finally, a number of other issues, such as expanded recreational needs, preservation and restoration of the environment, and even certain public health concerns, entail a geological component. Solutions to all of these social issues will involve geology, ensuring that knowledge of the geology of the Southwest will become increasingly important in the future.

Glossary

Accretionary prism (accretionary wedge). A generally wedge-shaped mass of tectonically deformed sediment at a subduction zone formed by the tectonic transfer of strata from the descending plate into the framework of the overriding plate (Jackson, 1997).

Agglomerate. A term used to describe pyroclastic rocks which contain an abundance of subangular or rounded fragments lying in a matrix of tuff. Agglomerates usually occur in or near volcanic vents or necks (Walton *et al.*, 1983).

Allochthonous. Formed or produced elsewhere than in the place now found. The term is widely applied to a mass of rock (as in a thrust sheet) overlying low-angle thrust faults in orogenic zones (Jackson, 1997). Antonym is **autochthonous.**

Alluvial. A term referring to unconsolidated subaerial sediments, especially young stream or river deposits.

Alluvial fan. A low, gently sloping mass of loose rock material, shaped like an open fan or a segment of a cone, deposited by a stream at the place where it issues from a narrow mountain valley onto a plain or broad valley. A fan is steepest near the mouth of the valley where its apex points upstream, and slopes gently and convexly outward with gradually decreasing gradient (Jackson, 1997). Fans are especially characteristic of arid or semiarid regions.

Anoxic. Lacking oxygen.

Argillite. Mudrock that has been subjected to low-grade metamorphism (Prothero and Schwab, 1996).

Arkose. A feldspar-rich sandstone, commonly coarse-grained and pink or reddish, that is typically composed of angular to subangular grains that may be either poorly or moderately well sorted. Arkose is usually derived from the rapid disintegration of granite or granitic rocks, and often closely resembles granite. Quartz is usually the dominant mineral, with feldspars constituting at least 25% (Jackson, 1997).

Asthenosphere. A weak layer in the Earth's upper mantle in which seismic waves are strongly attenuated. The asthenosphere is generally thought to contain a small amount of silicate melt and to be the source of magma, particularly beneath mid-ocean ridges. The asthenosphere underlies the lithosphere.

Autochthonous. Formed or produced in the place where now found. The term is widely applied to rocks that have not been displaced by faulting in orogenic zones (Jackson, 1997). Antonym is **allochthonous.**

Basement. A general term for the undifferentiated group of rocks that underlies the rocks of interest in an area. The term typically refers to igneous and metamorphic rocks of Precambrian age that underlie sedimentary strata.

Braided fluvial system. Braiding refers to repeated branching and rejoining of drainage channels to form an intricate pattern or network of small, interlacing stream channels. Thus, a **braided stream** or **braided river** (i.e. a **braided fluvial system**) is a drainage system characterized by braiding. Similarly, alluvial fans, alluvial plains, and deltas may be characterized by braided channels.

Claystone. Siliciclastic sedimentary rock in which clay-sized particles (< 1/256 in, or 0.0039 mm) dominate over silt-sized particles (1/16 to 1/256 in, or 0.0625 to 0.0039 mm) (Prothero and Schwab, 1996).

Diamictite. A poorly sorted detrital rock in which pebbles and larger grains float in a sandy or muddy matrix (Prothero and Schwab, 1996). Many diamictites are inferred to have a glacial origin.

Dolostone. Carbonate rock consisting of the mineral dolomite, $CaMg(CO_3)_2$.

Eolian. Pertaining to the wind, and to wind-laid deposits such as sand dunes.

Fanglomerate. Stream-channel conglomerate deposited on an alluvial fan (Prothero and Schwab, 1996).

Fenestrally laminated. Laminae (thin depositional layers) characterized by parallel, elongate open spaces.

Fluvial. Of or pertaining to rivers.

Foreland. Orogenic belts, which typically form on plate margins, are visualized from the perspective of the adjacent plate. Thus, the side closest to the margin, the front of the orogenic zone, is the **foreland**, the far side is the **hinterland**.

Foreland thrust belt. A linear zone in the foreland of an orogenic belt that is characterized by thrust faults.

Graded bedding. A type of bedding in which layers display a gradual and progressive change in particle size, usually from coarse at the base of the bed to fine at the top. Graded bedding may form under conditions in which the velocity of the prevailing current declined in a gradual manner, as by deposition from a swiftly moving, bottom-flowing current laden with suspended sediment (Jackson, 1997).

Graywacke. Or **wacke.** A siliclastic sedimentary rock composed of different proportions of quartz (typically 25–50%), feldspar, and rock fragments in a matrix of clay minerals and chlorite. Feldspar clasts are most often angular to subangular in shape. Grains of chert, mudrock, limestone, polycrystalline quartz, and volcanic rocks are common. The clay matrix constitutes 15% or more of the rock (Prothero and Schwab, 1996).

Greenstone. A general term applied to any dense, dark-green altered or metamorphosed mafic igneous rock. The color of greenstone generally results from the presence of chlorite, actinolite, and/or epidote (Jackson, 1997).

Hinterland. See **foreland.**

Hypabyssal. An igneous rock or intrusive body formed at a shallow depth in the crust (Jackson, 1997).

Intertidal. Referring to the sedimentary environment between low- and high-tide level. See **supratidal.**

Joint. A planar fracture, crack, or parting in a rock, without displacement across it (Jackson, 1997).

Karst. A type of topography formed on limestone, gypsum, and other soluble rocks, primarily by dissolution, that is characterized by sinkholes, caves, and underground drainage (Jackson, 1997).

Karstic. Adjective of **karst.**

Lacustrine. Pertaining to, produced by, or formed in a lake or lakes (Jackson, 1997).

Lithosphere. The rigid, outer layer of the Earth, comprising the crust and uppermost mantle. The lithosphere is typically 35–200 km thick, depending on tectonic setting. Continental lithosphere, especially in stable regions, is typically thicker than that of the oceans. The lithosphere is underlain by the asthenosphere.

Littoral. The region on the shore of the sea or of a large lake, or of or pertaining to the shore.

Marlstone. An indurated rock of the same composition as marl, a term loosely applied to an earthy deposit of clay and calcium carbonate formed under marine or freshwater conditions. Specifically, the term **marl** has been applied to newly formed deposits of shells mixed with clay (Jackson, 1997).

Migmatite. A composite rock found in medium- to high-grade metamorphic areas, composed of interlayered, quartz-rich igneous or igneous-appearing rock and metamorphic rock. The igneous rock may be a partial melt of the host metamorphic body.

Miogeocline. A wedge of shallow-water sediment that is built seaward at the continental margin (Jackson, 1997).

Mudrock. A general term for siliciclastic rock composed predominantly of silt-sized (1/16 to 1/256 in, or 0.0625 to 0.0039 mm) and clay-sized (< 1/256 in, or < 0.0039 mm) particles. **Mudrock** includes **siltstone, mudstone,** and **claystone.**

Mudstone. Referring to siliciclastic rocks, **mudstone** is indurated mud, which is a mixture of silt (particles 1/16 to 1/256 in, or 0.0625 to 0.0039 mm diameter) with between one-third and two-thirds clay (particles < 1/256 in, or < 0.0039 mm diameter). Referring to carbonate rocks, **mudstone** is limestone consisting mainly of lime mud (particle size < 0.03 mm), with less than 10% carbonate grains 0.03–2 mm in diameter. Grains are not in contact with each other. Compare **packstone** (Prothero and Schwab, 1996).

Mylonite. A fine-grained, foliated rock produced in ductile fault zones by the extreme granulation and shearing of rocks.

Orogen. A linear or arcuate region (**orogenic belt**) that has been subjected to folding and other deformation. Orogeny is the process by which structures within fold-belt mountainous areas were formed, including thrusting, folding, faulting, metamorphism, and intrusion of magma. In the Cenozoic, deformational structures (e.g.

faults and folds) and formation of the mountainous landscape are intimately related (Jackson, 1997).

Packstone. Limestone consisting dominantly of clastic grains of carbonate 0.03–2 mm in diameter. Grains are in contact with each other, and mud is present in interstices.

Paleosol. A soil that formed on a landscape in the past, with distinctive morphological features resulting from a soil-forming environment that no longer exists at the site (Jackson, 1997).

Pelagic. Referring to sediments deposited in deep ocean basins, at depths of from 4–6 km or more. The main constituents of pelagic sediments are clays derived from the continents, biogenic skeletal material of marine organisms, and minor amounts of minerals that form in place in deep-sea mud and ooze. Deep-sea sediments may also include volcanic ash, eolian dust, and meteoritic material from extraterrestrial impacts. The constituent particles settle out of the overlying water column (Prothero and Schwab, 1996).

Paludal. Pertaining to a marsh or swamp.

Pillow basalt. A term for basaltic rocks displaying pillow structure. Pillow structure is characterized by pillow-shaped masses (**pillows**) of basalt ranging in size from a few centimeters to a meter or more in cross section, and which are elongated in longitudinal section. Pillows may be close-fitting, the concavities of one fitting the convexities of adjacent pillows. The spaces between pillows are typically filled by scoriaceous material of basaltic composition, by altered basaltic glass, or by clastic sediments. Pillows are typically finer-grained toward the outsides, with glassy rinds. Pillow structure is inferred to be the result of subaqueous extrusion (Jackson, 1997).

Playa. The nearly flat, central part of a desert basin, in which water collects after a rain and evaporates to leave behind silt, clay, and evaporite deposits.

Pyroclastic. Pertaining to clastic (fragmental) material formed by volcanic explosion or aerial expulsion from a volcanic vent. See **tuff**.

Reactivation surface. An inclined bedding surface separating otherwise conformable cross-beds. It is formed by slight erosion of the lee side of a sand wave, megaripple, or bar, during a period when deposition is temporarily interrupted. Reactivation surfaces are abundant in sands deposited by the migration of megaripples or dunes in the tidal environment (Jackson, 1997).

Ripup clasts. Clasts derived as semiconsolidated mud deposits are ripped up by currents and transported to a new depositional site. Ripup clasts typically have a flat shape.

Selenite. The clear, colorless variety of the mineral gypsum, $CaSO_4 \cdot 2H_2O$. Selenite typically occurs in distinct, transparent monoclinic crystals or in large crystalline masses.

Shale. Any mudrock that exhibits lamination or fissility (i.e. can be easily split), or both.

Siliciclastic. A term referring to sedimentary rocks composed mainly of grains of quartz and other silicate minerals (in contrast with rocks composed of carbonate).

Siltstone. Siliciclastic sedimentary rock in which silt-sized particles (1/16 to 1/256 in, or 0.0625 to 0.0039 mm) comprise 50% or more of the rock (Prothero and Schwab, 1996).

Subaerial. Referring to conditions and processes that exist or operate in the open air on or immediately adjacent to the land surface, or to features and materials that are formed or situated on the land surface (Jackson, 1997).

Supracrustal. The term refers to stratified rocks, i.e. those formed at or near the Earth's surface.

Supratidal. Referring to the sedimentary environment above high-tide level, which is immersed only during unusually high spring tides or during storm surges (Prothero and Schwab, 1996). See **intertidal**.

Suture. A fault or fault zone that marks the boundary between two crustal blocks that were once widely separated, typically on different plates. The presence of a suture (or suture zone) implies that oceanic lithosphere once existed between the two blocks (Jackson, 1997).

Terrigenous. Derived from the land (Jackson, 1997).

Tetrapod. An animal with four limbs. An informal term to distinguish amphibians, reptiles, and mammals from aquatic classes in which paired limbs are absent or are fins (Jackson, 1997).

Transpression. In crustal deformation, a combination of strike–slip motion and compression perpendicular to the strike–slip zone.

Tuff. A general term for deposits of consolidated, fragmental material formed by volcanic explosion or aerial expulsion from a volcanic vent. Tuff deposits may be characterized as **ash-flow** or **fallout** (or **fall**). **Ash-flow tuffs** are deposits left from eruption clouds consisting of hot particles and gas, driven by gravity and moving laterally across the ground as density currents (pyroclastic flows). Many pyroclastic flows are generated by collapse of particle-laden eruption columns. Most pyroclastic flows move at high velocity downslope and along drainage systems, but some have enough volume and energy to move across hills and valleys. Pyroclastic flows are also known as ash flows, although the latter term is less accepted. **Fallout tuffs** are fragmental volcanic deposits composed of ash and pumice that is deposited primarily by settling under the influence of gravity from an eruption plume. Fallout deposits may be 'reworked' (i.e. suspended and redeposited) on the Earth's surface by wind or water.

Turbidite. A sediment or sedimentary rock deposited from a dense, swiftly moving, bottom-flowing current of water laden with suspended sediment. A turbidite is characterized by graded bedding, moderate sorting, and well-developed primary structures. Turbidite deposits range greatly in grain size, i.e. in the amount of sand vs. mud that characterize a given deposit. Turbidites are initially deposited in deep water, characterizing the deeper parts of sedimentary basins. The geometry of turbidite deposits can be highly complex, depending on the channel systems, or lack thereof, whereby the sediments were delivered to the slopes and floors of the basins (Jackson, 1997; Weimer and Slatt, 1999).

Varve. A cyclic sedimentary pair of laminae, as in certain shales and evaporite deposits, possibly deposited in a body of still water within one year's time.

Vergence. Direction of overturning or of inclination of a fold or fault. The direction toward which a structure is turned.

Volcaniclastic. Referring to sedimentary rocks rich in fragments of volcanic rock and/or glass (Prothero and Schwab, 1996).

Welding. With respect to volcanic tuffs, welding is the process by which tuff becomes indurated by the fusing together of glass shards under the combined action of heat retained by particles and interstitial gases, and of the weight of overlying material. The term **welded** refers to deposits having undergone welding.

References

Adams, J. W., Love, D. W., and Hawley, J. W., 1993, Third-day road log, from Carlsbad to Dark Canyon, Last Chance Canyon, Sitting Bull Falls, Rocky Arroyo and return to Carlsbad. *N. Mex. Geol. Soc. Guidebook* **44**, 69–86.

Algeo, T. J., and Seslavinsky, K. B., 1995, The Paleozoic world: Continental flooding, hypsometry, and sealevel. *Am. J. Sci.* **295**, 787–822.

Allen, P. A., Verlander, J. E., Burgess, P. M., and Audet, D. M., 2000, Jurassic giant erg deposits, flexure of the United States continental interior, and timing of the onset of Cordilleran shortening. *Geology* **28**, 159–62.

Alvarez, L. W., Alvarez, W., Azaro, F., and Michel, H. V., 1980, Extraterrestrial cause for the Cretaceous-Tertiary extinction. *Science* **208**, 1095–108.

Alvarez, W., Claeys, P., and Kieffer, S. W., 1995, Emplacement of Cretaceous-Tertiary boundary shocked quartz from Chicxulub crater. *Science* **269**, 930–5.

Alvarez, W., Staley, E., O'Connor, D., and Chan, M. A., 1998, Synsedimentary deformation in the Jurassic of southeastern Utah–A case of impact shaking? *Geology* **26**, 579–82.

Anderson, J. L., 1989, Proterozoic anorogenic granites of the southwestern United States. *In* Jenney, J. P., and Reynolds, S. J., eds., *Geologic Evolution of Arizona, Arizona Geological Society, Digest* **17**, 211–38.

Anderson, O. J., and Lucas, S. G., 1996, The base of the Morrison Formation (Upper Jurassic) of northwestern New Mexico and adjacent areas. *In* Morales, M., ed., *The Continental Jurassic, Mus. Northern Ariz. Bull.* **60**, 443–56.

Anderson, O. J., Lucas, S. G., Semken, S. C., Chenoweth, W. L., and Black, B. A., 1997, Third-day road log from Durango, Colorado, to Aztec, Farmington, and Shiprock, New Mexico. *N. Mex. Geol. Soc. Guidebook* **48**, 35–53.

Anderson, R. Y., 1993, The Castile as a "nonmarine" evaporite. *N. Mex. Geol. Soc. Guidebook* **44**, 12–13.

Arinobu, T., Ishiwatari, R., Kaiho, K., and Lamolda, M. A., 1999, Spike of pyrosynthetic polycyclic aromatic hydrocarbons associated with an abrupt decrease in

δ^{13}C of a terrestrial bimarker at the Cretaceous-Tertiary boundary at Caravaca, Spain. *Geology* **27**, 723–6.

Armstrong, A. K., 1995, Facies, diagenesis, and mineralogy of the Jurassic Todilto Limestone Member, Grants uranium district, New Mexico. *N. Mex. Bur. Mines Min. Res. Bull.* **153**, 41pp.

Atwater, T., and Stock, J., 1998, Pacific-North American plate tectonics of the Neogene southwestern United States: An update. *Int. Geol. Rev.* **40**, 375–402.

Austin, G. S., Barker, J. M., Crawford, J. E., Hawley, J. W., Love, D., Lucas, S. G., and Adams, J. W., 1993, First-day road log from Carlsbad to Whites City, Orla, Loving, Potash enclave and return to Carslbad. *N. Mex. Geol. Soc. Guidebook* **44**, 1–42.

Baars, D. L., 1974, Permian rocks of north-central New Mexico. *N. Mex. Geol. Soc. Guidebook* **25**, 167–9.

    1983, *The Colorado Plateau.* Albuquerque, NM: University of New Mexico Press, 279pp.

    1988, Paradox basin. *In* Sloss, L. L., *The Geology of North America*, vol. D-2, *Sedimentary Cover – North American Craton*, pp. 114–122. Boulder, CO: The Geological Society of North America.

Baldridge, W. S., Ferguson, J. F., Braile, L. W., Wang, B., Eckhardt, K., Evans, D., Schultz, C., Gilpin, B., Jiracek, G. R., and Biehler, S., 1994, The western margin of the Rio Grande rift in northern New Mexico: An aborted boundary? *Geol. Soc. Am. Bull.* **105**, 1538–51.

Baldridge, W. S., Keller, G. R., Haak, V., Wendlandt, E., Jiracek, G. R., and Olsen, K. H., 1995, The Rio Grande rift. *In* Olsen, K. H., ed., *Continental Rifts: Evolution, Structure, Tectonics*, pp. 233–75. Developments in Geotectonics. Amsterdam: Elsevier.

Barker, D. S., 1979, Cenozoic magmatism in the Trans-Pecos province: relation to the Rio Grande rift. *In* Riecker, R. E., ed., *Rio Grande Rift: Tectonics and Magmatism.* Am. Geophys. U., Spec. Publ., pp. 382–92. Washington, D.C.: American Geophysical Union.

Barth, A. P., Tosdal, R. M., Wooden, J. L., and Howard, K. A., 1997, Triassic plutonism in southern California: Southward younging of arc initiation along a truncated continental margin. *Tectonics* **16**, 290–304.

Bebout, D. G., and Kerans, C., eds., 1993, Guide to the Permian reef geology trail, McKittrick Canyon, Guadalupe Mountains Nat. Park. *Texas Bur. Econ. Geology, Guidebook* **26**, 48pp.

Beus, S. S., 1990, The Temple Butte Formation. *In* Beus, S. S., and Morales, M., eds., *Grand Canyon Geology*, pp. 107–17. Oxford University Press.

Beus, S. S., and Morales, M., 1990, Introducing the Grand Canyon. *In* Beus, S. S., and Morales, M., eds., *Grand Canyon Geology*, pp. 1–9. Oxford University Press.

Bickford, M. E., Soegaard, K., Nielsen, K. C., and McLelland, J. M., 2000, Geology and geochronology of Grenville-age rocks in the Van Horn and Franklin Mountains area, west Texas: Implications for the tectonic evolution of Laurentia during the Grenville. *Geol. Soc. Am. Bull.* **112**, 1134–48.

Bilodeau, W. L., 1982, Tectonic models for Early Cretaceous rifting in southwestern Arizona. *Geology* **10**, 466–70.

Blakey, R. C., 1989, Triassic and Jurassic geology of the southern Colorado Plateau. *In* Jenney, J. P., and Reynolds, S. J., Geologic evolution of Arizona, pp. 369–96. Tucson, AZ: Arizona Geological Society.

1990a, Supai Group and Hermit Formation. *In* Bues, S. S., and Morales, M., eds., *Grand Canyon Geology*, pp. 147–82. Oxford University Press.

1990b, Stratigraphy and geologic history of Pennsylvanian and Permian rocks, Mogollon Rim region, central Arizona and vicinity. *Geol. Soc. Am. Bull.* **102**, 1189–217.

1996, Jurassic regional unconformities of the Colorado Plateau: The key to accurate correlation and interpretation of the area's Jurassic geologic history. *In* Morales, M., ed., *The Continental Jurassic, Museum Northern Ariz. Bull.* **60**, 439–41.

Blakey, R. C., and Knepp, R., 1989, Pennsylvanian and Permian geology of Arizona. *In* Jenney, J. P., and Reynolds, S. J., eds., *Geol. Evolution of Arizona, Arizona Geol. Soc. Digest.* **17**, 313–47.

Blakey, R. C., and Middleton, L. T., 1986, Triassic-Jurassic continental systems, northern Arizona. *In* Nations, J. D., Conway, C. M., and Swann, G. A., eds., Geology of central and northern Arizona, *Geol. Soc. Am, Rocky Mtn. Sec. Guidebook, Flagstaff, AZ*, pp. 93–110.

Blakey, R. C., Havholm, K. G., and Jones, L. S., 1996, Stratigraphic analysis of eolian interactions with marine and fluvial deposits, Middle Jurassic Page Sandstone and Carmel Formation, Colorado Plateau, U. S. A. *J. Sed. Res.* **66**, 324–42.

Bland, D. M., 1992, Coalbed methane from the Fruitland Formation, San Juan Basin, New Mexico. *N. Mex. Geol. Soc. Guidebook* **43**, 373–83.

Blount, J. G., 1983, The geology of the Rancho Los Filtros area, Chihuahua, Mexico. *In* Clark, K. F., and Goodell, P. C., eds., *Geology and Mineral Resources of North-Central Chihuahua*, pp. 157–64. El Paso: El Paso Geological Society.

Bond, G. C., Nickeson, P., and Kominz, M. A., 1984, Breakup of a supercontinent between 625 Ma and 555 Ma: New evidence and implications for continental histories. *Earth Planet. Sci. Lett.* **70**, 325–45.

Bowring, S. A., and Erwin, D. H., 1998, A new look at evolutionary rates in deep time: Uniting paleontology and high-precision geochronology. *GSA Today* **8**(9), 1–8.

Bowring, S. A., and Karlstrom, K. E., 1990, Growth, stabilization, and reactivation of Proterozoic lithosphere in the southwestern United States. *Geology* **18**, 1203–6.

Bowring, S. A., Williams, I. S., and Compston, W., 1989, 3.96 Ga gneisses from the Slave Province, Northwest Territories, Canada. *Geology* **17**, 971–5.

Bowring, S. A., Erwin, D. H., Jin, Y. G., Martin, M. W., Davidek, K., and Wang, W., 1998, U/Pb zircon geochronology and tempo of the end-Permian mass extinction. *Science* **280**, 1039–45.

Bruno, L., and Chafetz, H. S., 1988, Depositional environment of the Cable Canyon Sandstone; a Mid-Ordovician sandwave complex from southern New Mexico. *N. Mex. Geol. Soc. Guidebook* **39**, 127–34.

Budnik, R. T., 1986, Left-lateral intraplate deformation along the Ancestral Rocky Mountains: implications for late Paleozoic plate motions. *Tectonophysics* **132**, 195–214.

Burchfiel, B. C., and Davis, G. A., 1972, Structural framework and evolution of the southern part of the Cordilleran orogen, western United States. *Am. J. Sci.* **272**, 97–118.

1975, Nature and controls of Cordilleran orogenesis, western United States: Extensions of an earlier synthesis. *Am. J. Sci.* **275**-A, 363–96.

Burchfiel, B. C., and Royden, L. H., 1991, Antler orogeny: A Mediterranean-type orogeny. *Geology* **19**, 66–9.

Burchfiel, B. C., Cowan, D. S., and Davis, G. A., 1992, Tectonic overview of the Cordilleran orogen in the western United States. *In* Burchfiel, B. C., Lipman, P. W., and Zoback, M. L., eds., *The Cordilleran Orogen: Conterminous U. S.*, vol. G-3, pp. 407–79. Boulder, CO: Geological Society of North America.

Burgess, P. M., Gurnis, M., and Moresi, L., 1997, Formation of sequences in the cratonic interior of North America by interaction between mantle, eustatic, and stratigraphic processes. *Geol. Soc. Am. Bull.* **108**, 1515–35.

Burrett, C., and Berry, R., 2000, Proterozoic Australia-Western United States (AUSWUS) fit between Laurentia and Australia. *Geology* **28**, 103–6.

Busby-Spera, C. J., 1988, Speculative tectonic model for the early Mesozoic arc of the southwest Cordilleran United States. *Geology* **16**, 1121–5.

Catuneanu, O., Beaumont, C., and Waschbusch, P., 1997, Interplay of static loads and subduction dynamics in foreland basins: Reciprocal stratigraphies and the "missing" peripheral bulge. *Geology* **25**, 1087–90.

Chamberlin, R. M., 1983, Cenozoic domino-style crustal extension in the Lemitar Mountains, New Mexico: A summary. *N. Mex. Geol. Soc. Guidebook* **34**, 111–18.

Chan, M. A., 1989, Erg margin of the Permian White Rim Sandstone, SE Utah. *Sedimentology* **36**, 235–51.

Chapin, C. E., and Cather, S. M., 1994, Tectonic setting of the axial basins of the northern and central Rio Grande rift. *In* Keller, G. R., and Cather, S. M., eds., *Basins of the Rio Grande Rift: Structure, Stratigraphy, and Tectonic Setting*, pp. 5–25. Boulder, CO: Geological Society of America Special Paper **291**.

Chapin, C. E., and Seager, W. R., 1975, Evolution of the Rio Grande rift in the Socorro and Las Cruces areas. *N. Mex. Geol. Soc. Guidebook* **26**, 297–321.

Christeson, G. L., Nakamura, Y., Buffler, R. T., Morgan, J., and Warner, M., 2001, Deep crustal structure of the Chicxulub impact crater. *Geophys Res.* **106**, 21, 751–21, 769.

Christiansen, R. L., and Yeats, R. S., 1992, Post-Laramide geology of the U. S. Cordilleran region. *In* Burchfiel, B. C., Lipman, P. W., and Zoback, M. L., eds, *The Geology of North America*, vol. G-3, *The Cordilleran Orogen: Conterminous U. S.* pp. 261–406. Boulder, CO: Geological Society of America.

Clark, K. F., Foster, C. T., and Damon, P. E., 1982, Cenozoic mineral deposits and subduction-related magmatic arcs in Mexico. *Geol. Soc. Am. Bull.* **93**, 533–44.

Clemons, R. E., 1998, Geology of the Florida Mountains, southwestern New Mexico, *N. Mex. Bur. Mines Miner. Res., Mem.* **43**, 112pp.

Clemons, R. E., and Osburn, G. R., 1986, Geology of the Truth or Consequences area. *N. Mex. Geol. Soc. Guidebook* **37**, 69–81.

Colbert, E. H., 1995, *The Little Dinosaurs of Ghost Ranch.* New York: Columbia University Press, 250p.

Colpitts, R. M., Jr., 1989, Permian reference section for southeastern Zuni Mountains, Cibola County, New Mexico. *N. Mex. Geol. Soc. Guidebook* **40**, 177–81.

Condie, K. C., 1981, Precambrian rocks of the southwestern United States and adjacent areas of Mexico, *N. Mex. Bur Mines Min. Res., Resource Map* 13.

Condon, S. M., 1997, Geology of the Pennsylvanian and Permian Cutler Group and Permian Kaibab Limestone in the Paradox basin, southeastern Utah and southwestern Colorado. *U. S. Geol. Surv. Bulletin*, 2000-P, 37pp.

Conway, C. M., and Silver, L, T., 1989, Early Proterozoic rocks (1710-1615 Ma) in central to southeastern Arizona. *In* Jenney, J. P., and Reynolds, S. J., *Geologic evolution of Arizona, Ariz. Geol. Soc. Digest* **17**, 165–86.

Conway, C. M., Karlstrom, K. E., Silver, L. T., and Wrucke, C. T., 1987, Tectonic and magmatic contrasts across a two-province Proterozoic boundary in central Arizona. *In Geologic Diversity of Arizona and its Margins: Excursions to Choice Areas, Ariz. Bur. Geol. Mineral Technol. Spec. Paper* 5, 158–75.

Corsetti, F. A., and Hagadorn, J., 2000, Precambrian–Cambrian transition: Death Valley, United States. *Geology* **28**, 299–302.

Cotkin, S. J., 1992, Ordovician-Silurian tectonism in northern California: The Callahan event. *Geology* **20**, 821–4.

Cotkin, S. J., Cotkin, M. L., and Armstrong, R. L., 1992, Early Paleozoic blueschist from the Schist of Skookum Gulch, eastern Klamath Mountains, northern California. *J. Geol.* **100**, 323–38.

Cowan, D. S., and Bruhn, R. L., 1992, Late Jurassic to early Late Cretaceous geology of the U. S. Cordillera. *In* Burchfiel, B. C., Lipman, P. W., and Zoback, M. L., eds., pp. 169–203. *The Geology of North America*, vol. G-3, *The Cordilleran Orogen: Conterminous U. S.*, pp. 169–203. Boulder, CO: Geological Society of America.

Cygan, R. T., Crossey, L. J., and Marin, L. E., 1996, Researchers focus on Earth's response to hypervelocity impacts. *Eos, Trans. Am. Geophys. Union* **77**, 197, 199.

Dalrymple, G. B., 1991, *The Age of the Earth*. Stanford, CA: Stanford University Press. 474pp.

Dalziel, I. W. D., 1997, Neoproterozoic-Paleozoic geography and tectonics: Review, hypothesis, environmental speculation. *Geol. Soc. Am.* **109**, 16–42.

Dalziel, I. W. D., and McMenamin, M. A. S., 1995, Are Neoproterozoic glacial deposits preserved on the margins of Laurentia related to the fragmentation of two supercontinents? Comment. *Geology* **23**, 959–60.

Daniel, C. G., Karlstrom, K. E., Williams, M. L., and Pedrick, J. N., 1995, The reconstruction of a Middle Proterozoic orogenic belt in north-central New Mexico, U.S.A. *N. Mex. Geol. Soc. Guidebook* **46**, 193–200.

Dann, J. C., 1991, Early Proterozoic ophiolite, central Arizona. *Geology* **19**, 590–3.
   1997, Pseudostratigraphy and origin of the Early Proterozoic Payson ophiolite, central Arizona. *Geol. Soc. Am. Bull.* **109**, 347–65.

DeCelles, P. G., and Mitra, G., 1995, History of the Sevier orogenic wedge in terms of critical taper models, northeast Utah and southwest Wyoming. *Geol. Soc. Am. Bull.* **107**, 454–62.

DeCelles, P. G., Lawton, T. F., and Mitra, G., 1995, Thrust timing, growth of structural culminations, and synorogenic sedimentation in the type Sevier orogenic belt, western United States. *Geology* **23**, 699–702.

Detrick, R., Collins, J., Stephen, R., and Swift, S., 1994, In-situ evidence for the nature of the seismic layer 2/3 boundary in oceanic-crust. *Nature* **370**, 288–90.

Dickinson, W. R., 1981, Plate tectonic evolution of the southern Cordillera. *Arizona Geol. Soc. Digest* **14**, 113–35.

Dickinson, W. R., and Snyder, W. S., 1978, Plate tectonics of the Laramide orogeny. *Geol. Soc. Amer. Memoir* **151**, 355–66.

DuBois, S. M., and Smith, A. W., 1980, The 1887 Earthquake in San Bernardino Valley, Sonora: Historic accounts and intensity patterns in Arizona. *Arizona Bureau of Geology and Mineral Technology, Spec. Paper* No. 3, 112pp.

du Bray, E. A., and Pallister, J. S., 1999, Recrystallization and anatexis along the plutonic-volcanic contact of the Turkey Creek caldera, Arizona. *Geol. Soc. Am. Bull.* **111**, 143–53.

Duebendorfer, E. M., and Christensen, C., 1995, Synkinematic(?) intrusion of the 'anorogenic' 1425 Ma Beer Bottle Pass pluton, southern Nevada. *Tectonics* **14**, 168–84.

Duebendorfer, E. M., Chamberlain, K. R., and Jones, C. S., 2001, Paleoproterozoic tectonic history of the Cerbat Mountains, northwestern Arizona: Implications for crustal assembly in the southwestern United States. *Geol. Soc. Am. Bull.* **113**, 575–90.

Eaton, G. P., 1979, Regional geophysics, Cenozoic tectonics, and geologic resources of the Basin and Range province and adjoining regions. *In* Newman, G. W., and Goode, H. D., eds., *1979 Basin and Range Sympos.*, pp. 11–39. Denver, CO: Rocky Mountain Association of Geologists and Utah Geological Association.

1986, A tectonic redefinition of the southern Rocky Mountains, *Tectonophysics* **132**, 163–93.

Elston, D. P., 1989, Middle and Late Proterozoic Grand Canyon Supergroup, Arizona. *In* Elston, D. P., Billingsley, G. H., and Young, R. A., eds., *Geology of Grand Canyon, Northern Arizona (with Colorado River Guides)*, 28th Int. Geol. Congress, Field Trips T115, T315, pp. 94–105. Washington, D.C.: American Geophysical Union.

1993, Middle and Early-Late Proterozoic Grand Canyon Supergroup, northern Arizona. *In* Reed, J. C., Jr., Bickford, M. E., Houston, R. S., Link, P. K., Rankin, D. W., Sims, P. K., and Van Schmus, W. R., eds., *The Geology of North America*, Vol. C-2, *Precambrian: Conterminous U. S.*, pp. 521–9. Boulder, CO: Geological Society of America.

Elston, D. P., Link, P. K., Winston, D., and Horodyski, R. J., 1993, Correlations of Middle and Late Proterozoic successions. *In* Reed, J. C., Jr., Bickford, M. E., Houston, R. S., Link, P. K., Rankin, D. W., Sims, P. K., and Van Schmus, W. R., eds., *The Geology of North America*, vol. C-2, *Precambrian: Conterminous U. S.*, pp. 468–595. Boulder, CO: Geological Society of America.

Elston, W. P., 1984, Subduction of young oceanic lithosphere and extensional orogeny in southwestern North America during mid-Tertiary time. *Tectonics* **3**, 229–50.

Ericksen, M. C., and Slingerland, R., 1990, Numerical simulations of tidal and wind-driven circulation in the Cretaceous Interior Seaway of North America. *Geol. Soc. Am. Bull.* **102**, 1499–516.

Erwin, D. H., 1994, The Permo-Triassic extinction. *Nature* **367**, 231–6.

Eshet, Y., Rampino, M. R., and Visscher, H., 1995, Fungal event and palynological record of ecological crisis and recovery across the Permian-Triassic boundary. *Geology* **23**, 976–70.

Fagerstrom, J. A., and Weidlich, O., 1999, Origin of the upper Capitan-Massive limestone (Permian), Guadalupe Mountains, New Mexico–Texas: Is it a reef? *Geol. Soc. Am. Bull.* **111**, 159–76.

Fenneman, N. M., 1928, Physiographic divisions of the United States. *Ann. Assoc. Am. Geogr.*, 3rd edn., **18**, 261–353.

Ferrari, L., López-Martinez, M., Aguirre-Díaz, G., and Carrasco-Núñez, G., 1999, Space-time patterns of Cenozoic arc volcanism in central Mexico: from the Sierra Madre Occidental to the Mexican volcanic belt. *Geology* **27**, 303–6.

Fialko, Y., Simons, M., and Khazan, Y., 2001, Finite source modelling of magmatic unrest in Socorro, New Mexico, and Long Valley, California. *Geophys. J. Int.* **146**, 191–200.

Fillmore, R., 2000, *The Geology of the Parks, Monuments, and Wildlands of Southern Utah.* University of Utah Press, 268pp.

Flesch, L. M., Holt, W. E., Haines, A. J., and Shen-Tu, B., 2000, Dynamics of the Pacific-North American plate boundary in the western United States. *Science* **287**, 834–6.

Ford, T. D., 1990, Chapter 4: Grand Canyon Supergroup: Nankoweap Formation, Chuar Group, and Sixtymile Formation. *In* Beus, S. S., and Morales, M., eds., *Grand Canyon Geology*, pp. 49–70. Oxford University Press/Museum of Northern Arizona.

Fountain, D. M., and Christensen, N. I., 1989, Composition of the continental crust and upper mantle; a review. *In* Pakiser, I. C., and Mooney, W. D., eds., *Geophysical Framework of the Continental United States. Geol. Soc. of Am. Mem.* **172**, 711–42.

Froude, D. O., Ireland, T. R., Kinney, P. D., Williams, R. S., Compston, W., Williams, A. R., and Myers, J.S., 1983, Ion microprobe identification of 4,100–4,200 Myr-old detrital zircons. *Nature* **304**, 616–18.

Giles, K. A., 1998, The allochthonous nature of Lower Mississippian Waulsortian mounds in the Sacramento Mountains, New Mexico. *N. Mex. Geol. Soc. Guidebook* **49**, 155–60.

Glazner, A. F., and Ussler, W., III, 1988, Trapping of magma at midcrustal density discontinuities. *Geophys. Res. Lett.* **15**, 673–5.

Goldhammer, R. K., Lehmann, P. J., and Dunn, P. A., 1993, The origin of high-frequency platform carbonate cycles and third-order sequences (Lower Ordovician El Paso Gp, west Texas): Constraints from outcrop data and stratigraphic modeling. *J. Sed. Pet.* **63**, 318–59.

Goscombe, B. D., and Everard, J. L., 1999, Macquarie Island mapping reveals three tectonic phases. *Eos, Trans. Am. Geophys. Union* **80**, 50, 55.

Grambling, J. A., Williams, M. L., and Mawer, C. K., 1988, Proterozoic tectonic assembly of New Mexico. *Geology* **16**, 724–7.

Grand, S. P., van der Hilst, R. D., and Widiyantoro, S., 1997, Global seismic tomography: A snapshot of convection in the Earth. *GSA Today* **7**(4), 1–7.

Grant, K., and Owen, D. E., 1974, The Dakota Sandstone (Cretaceous) of the southern part of the Chama basin, New Mexico – A preliminary report on its stratigraphy, paleontology, and sedimentology. *N. Mex. Geol. Soc. Guidebook* **25**, 239–9.

Gupta, S. C., Ahrens, T. J., and Yang, W., 2001, Shock-induced vaporization of anhydrite and global cooling from the K/T impact. *Science* **188**, 399–412.

Gurnis, M., 1992, Long-term controls on eustatic and epeirogenic motions by mantle convection. *GSA Today* **2**, 141, 144–5, 156–7.

Hames, W. E., Renne, P. R., and Rupple, C., 2000, New evidence for geologically instantaneous emplacement of earliest Jurassic central Atlantic magmatic province basalts on the North American margin. *Geology* **28**, 859–62.

Handschy, J. W., and Dyer, R., 1987, Polyphase deformation in Sierra del Cuervo, Chihuahua, Mexico: Evidence for Ancestral Rocky Mountain tectonics in the Ouachita foreland of northern Mexico. *Geol. Soc. Am. Bull.* **99**, 618–32.

Handschy, J. W., Keller, G. R., and Smith, K. J., 1987, The Ouachita system in northern Mexico. *Tectonics* **6**, 323–30.

Haq, B. U., and Van Eysinga, F. W. B., 1987, *Geological Time Table*, Fourth revised edition. Amsterdam: Elsevier Science.

Harriet, M., 1992, Conodont biostratigraphy and paleoenvironment of the Surprise Canyon Formation (Late Mississippian), Grand Canyon, Arizona. M. S. Thesis, University of Northern Arizona, Flagstaff.

Harris, W. H., and Levey, J. S., 1975, *New Columbia Encyclopedia*. New York and London: Columbia University Press. 3052pp.

Hawkins, D. P., Bowring, S. A., Ilg, B. R., Karlstrom, K. E., and Williams, M. L., 1996, U-Pb geochronologic constraints on the Paleoproterozoic crustal evolution of the upper Granite Gorge, Grand Canyon, Arizona. *Geol. Soc. Am. Bull.* **108**, 1167–81.

Heiken, G., Goff, F., Gardner, J. N., and Baldridge, W. S., 1990, The Valles/Toledo caldera complex, Jemez volcanic field, New Mexico. *A. Rev. Earth Planet. Sci.* **18**, 27–53.

Heller, P. L., and Angevine, C. L., 1985, Sea-level cycles during the growth of Atlantic-type oceans. *Earth Planet. Sci. Lett.* **75**, 417–26.

Heller, P. L., Anderson, D. L., and Angevine, C. L., 1996, Is the middle Cretaceous pulse of rapid sea-floor spreading real or necessary? *Geology* **24**, 491–4.

Hendricks, J. D., and Stevenson, G. M., 1990, Grand Canyon Supergroup: Unkar Group. *In* Beus, S. S., and Morales, M., *Grand Canyon Geology*, Oxford University Press/Museum of Northern Arizona.

Henry, C. D., and Aranda-Gomez, J. J., 1992, The real southern Basin and Range: Mid- to late Cenozoic extension in Mexico. *Geology* **20**, 701–4.

Henry, C. D., and Price, J. G., 1986, The Van Horn Mountains caldera, Trans-Pecos Texas: geology and development of a small (10-km^2) ash-flow caldera. *Texas Bur. Econ. Geol., Report of Invest.* No. 151, 46pp.

Heymann, D., Yancey, T. E., Wolbach, W. S., Thiemens, M. H., Johnson, E. A., Roach, D., and Moecker, S., 1998, Geochemical markers of the Cretaceous-Tertiary boundary event at Brazos River, Texas, USA. *Geochim. Cosmochim. Acta* **62**, 173–81.

Hintze, L. F., 1988, *Geologic History of Utah*. Brigham Young University Geologic Studies, Spec. Publ. 7, 202pp.

Hoffman, P. F., 1989, Precambrian geology and tectonic history of North America. *In* Bally, A. W., and Palmer, A. R., eds., *The Geology of North America*, v. A, *The Geology of North America; An Overview*, pp. 447–512. Boulder, CO: Geological Society of America.

Holbrook, W. S., Lizarralde, D., McGeary, S., Bangs, N., and Diebold, J., 1999, Structure and composition of the Aleutian island arc and implications for continental crustal growth. *Geology* **27**, 31–4.

Hopkins, R. L., 1990, Chap. 12, In Beus, S. S., and Morales, M., eds., *Grand Canyon Geology*, pp. 225–45. Oxford University Press.

Hough, R. M., Gilmour, I., Pillinger, C. T., Langenhorst, F., and Montanari, A., 1997, Diamonds from the iridium-rich K-T boundary layer at Arroyo el Mimbral, Tamaulipas, Mexico. *Geology* **25**, 1019-22.

Humphreys, E. D., 1995, Post-Laramide removal of the Farallon slab, western United States. *Geology* **23**, 987–90.

Huntoon, J. E., and Chan, M. A., 1987, Marine origin of paleotopographic relief on eolian White Rim Sandstone (Permian), Elaterite basin, Utah. *Am. Assoc. Petrol. Geol. Bull.* **71**, 1035–45.

Idnum, M., and Giddings, J. W., 1995, Paleoproterozoic-Neoproterozoic North America-Australia link: New evidence from paleomagnetism. *Geology* **23**, 149–52.

Ingersoll, R. V., 1997, Phanerozoic tectonic evolution of central California and environs. *Int. Geol. Rev.* **39**, 957–72.

2000, Models for origin and emplacement of Jurassic ophiolites of northern California. *Geol. Soc. Am., Spec. Paper* **349**, 395–402.

Irwin, J. H., Stevens, P. R., and Cooley, M. E., 1971, Geology of the Paleozoic rocks, Navajo and Hopi Indian reservations, Arizona, New Mexico, and Utah. *U. S. Geol. Survey, Prof. Paper* **521**-C, 32pp.

Izett, G. A., Cobban, W. A., Dalrymple, G. B., and Obradovich, J. D., 1998, ^{40}Ar/^{39}Ar age of the Manson impact structure, Iowa, and correlative impact ejecta in the Crow Creek Member of the Pierre Shale (Upper Cretaceous), South Dakota and Nebraska. *Geol. Soc. Am. Bull.* **110**, 361–76.

Jablonski, D., and Raup, D. M, 1995, Selectivity of end-Cretaceous marine bivalve extinctions. *Science* **268**, 389–91.

Jackson, J.A., ed., 1997, *Glossary of Geology*, fourth edition. Alexandria, VA: American Geological Institute. 769pp.

Jackson, M. P. A., Schultz-Ela, D. D., Hudec, M. R., Watson, I. A., and Porter, M. L., 1998, Structure and evolution of Upheaval Dome: A pinched-off salt diapir. *Geol. Soc. Am. Bull.* **110**, 1547–73.

Jagnow, D. H., and Jagnow, R. R., 1992, *Stories from Stones. The Geology of the Guadalupe Mountains.* Carlsbad, NM: Carlsbad Caverns-Guadalupe Mountains Association, 41pp.

Jansma, P. E., and Speed, R. C., 1995, Kinematics of underthrusting in the Paleozoic Antler foreland basin, *J. Geol.* **103**, 559–75.

Jones, C. H., Unruh, J. R., and Sonder, L. J., 1996, The role of gravitational potential energy in active deformation in the southwestern United States. *Nature* **381**, 37–41.

Jordan, T. E., Isacks, B. L., Allmendinger, R. W., Brewer, J. A., Ramos, V. A., and Ando, C. J., 1983, Andean tectonics related to geometry of subducted Nazca plate. *Geol. Soc. Am. Bull.* **94**, 341–61.

Kaiho, K., Kajiwara, Y., Nakano, T., Miura, Y., Kawahata, H., Tazaki, K., Ueshima, M., Chen, Z. Q., and Shi, G. R., 2001, End-Permian catastrophe by a bolide impact: Evidence of a gigantic release of sulfur from the mantle. *Geology* **29**, 815–18.

Karlstrom, K. E., 1998, Introduction to special issues: Lithospheric structure and evolution of the Rocky Mountains (Parts I and II). *Rocky Mountain Geology* **33**, 157–9.

1999, Introduction to special issues, Part II: Nature of tectonic boundaries in lithosphere of the Rocky Mountains. *Rocky Mountain Geology* **34**, 1–4.

Karlstrom, K. E., and Bowring, S. A., 1988, Early Proterozoic assembly of tectonostratigraphic terranes in southwestern North America. *J. Geol.* **96**, 561–76.

1993, Proterozoic orogenic history of Arizona. *In* Reed, J. C., Jr., Bickford, M. E., Houston, R. S., Link, P. K., Rankin, D. W., Sims, P. K., and Van Schmus, W. R., eds., *The Geology of North America, v. C-2, Precambrian: Conterminous U. S.*, pp. 188–211. Boulder, CO: Geological Society of America.

Karlstrom, K. E., and Humphreys, E. D., 1998, Persistent influence of Proterozoic accretionary boundaries in the tectonic evolution of southwestern North America: Interaction of cratonic grain and mantle modification events. *Rocky Mountain Geology* **33**, 161–79.

Karlstrom, K. E., Williams, M. L., McLelland, J., Geissman, J. W., and Åhäll, K.-I., 1999, Refining Rodinia: Geologic evidence for the Australia-western U. S. connection in the Proterozoic. *GSA Today* **9**(10), 1–7.

Karlstrom, K. E., Åhäll, K.-I., Harlan, S. S., Williams, M. L., McLelland, J., and Geissman, J. W., 2001, Long-lived (1.8-1.0 Ga) convergent orogen in southern Laurentia, its extensions to Australia and Baltica, and implications for refining Rodinia. *Precambrian Res.* **111**, 5–30.

Kay, S. M., and Mpodozis, C., 2001, Central Andean ore deposits linked to evolving shallow subduction systems and thickening crust. *GSA Today* **11**(3), 4–9.

Keller, P. C., Bockoven, N. T., and McDowell, F. W., 1982, Tertiary volcanic history of the Sierra del Gallego area, Chihuahua, Mexico. *Geol. Soc. Am. Bull.* **93**, 303–14.

Kelley, S. A., and Chapin, C. E., 1995, Apatite fission-track thermochronology of southern Rocky Mountain-Rio Grande rift-western High Plains provinces. *N. Mex. Geol. Soc. Guidebook* **46**, 87–96.

Kenny, R., and Knauth, L. P., 2001, Stable isotope variations in the Neoproterozoic Beck Spring Dolomite and Mesoproterozoic Mescal Limestone paleokarst: Implications for life on land in the Precambrian. *Geol. Soc. Am. Bull.* **113**, 650–8.

Kirkland, D. W., Denison, R. E., and Evans, R., 1995, Middle Jurassic Todilto Formation of northern New Mexico and southwestern Colorado: Marine or nonmarine? *N. Mex. Bur. Mines Min. Res. Bull.* **147**, 37pp.

Kocurek, G., 1981, Erg reconstruction: the Entrada Sandstone (Jurassic) of northern Utah and Colorado. *Palaeogeogr. Palaeoclimatol. Palaeoecol.* **36**, 125–53.

Koeberl, C., Gilmour, I., Reimold, W. U., Claeys, P., and Ivanov, B., 2002, End-Permian catastrophe by bolide impact: Evidence of a gigantic release of sulfur from the mantle: Comment and Reply. *Geology* **30**, 855–6.

Kowallis, B. J., Christiansen, E. H., Deino, A. L., Zhang. C. N., and Everett, B. H., 2001, The record of Middle Jurassic volcanism in the Carmel and Temple Cap Formations of southwestern Utah. *Geol. Soc. Am. Bull.* **113**, 373–87.

Kriens, B. J., Shoemaker, E. M., and Herkenhoff, K. E., 1999, Geology of the Upheaval Dome impact structure, southeast Utah. *J. Geophys. Res.* **104** (E8), 18 867–87.

Kues, B. S., 2001, The Pennsylvanian System in New Mexico – overview with suggestions for revision of stratigraphic nomenclature. *N. Mex. Geol.* **23**, 103–22.

Kyle, J. R., ed., 2000, Geology and historical mining, Llano uplift region, central Texas. *Guideb.* **20**, *Austin Geol. Soc.*, 111pp.

Lallemand, S. E., and Tsien, H.-H., 1997, An introduction to active collision in Taiwan. *Tectonophysics* **274**, 1–4.

Larson, E. E., Patterson, P. E., and Mutschler, F. E., 1994, Lithology, chemistry, age, and origin of the Proterozoic Cardenas Basalt, Grand Canyon, Arizona. *Precambrian Res.* **65**, 255–76.

Larson, R. L., 1991, Latest pulse of the Earth: Evidence for a mid-Cretaceous superplume. *Geology* **19**, 547–50.

Laughlin, A. W., Poths, J., Healey, H. A., Reneau, S., and WoldeGabriel, G., 1994, Dating of Quaternary basalts using the cosmogenic ^{3}He and ^{14}C methods with implications for excess ^{40}Ar. *Geology* **22**, 135–8.

Lawton, T. F., and McMillan, N. J., 1999, Arc abandonment as a cause for passive continental rifting: Comparison of the Jurassic Mexican Borderland rift and the Cenozoic Rio Grande rift. *Geology* **27**, 779–82.

Lawton, T. F., McMillan, N. J., McLemore, V. T., and Hawley, J. W., 2000, Second-day road log, from Lordsburg to Deming via Little Hatchet Mountains and Victorio Mountains. *N. Mex. Geol. Soc. Guidebook* **51**, 17–30.

Leckie, R. M., Kirkland, J. I., and Elder, W. P., 1997, Stratigraphic framework and correlation of a principal reference section of the Mancos Shale (Upper Cretaceous), Mesa Verde, Colorado. *N. Mex. Geol. Soc. Guidebook* **48**, 163–216.

Lerner-Lam, A. L., Sheehan, A., Grand, S., Humphreys, E., Dueker, K., Hessler, E., Guo, H., Lee, D.-K., and Savage, M., 1998, Deep structure beneath the southern Rocky Mountains from the Rocky Mountain Front Broadband Seismic Experiment. *Rocky Mountain Geology* **33**, 199–216.

Leroux, H., Warme, J. E., and Doukhan, J.-C., 1995, Shocked quartz in the Alamo breccia, southern Nevada: Evidence for a Devonian impact event. *Geology* **23**, 1003–6.

Levin, H. L., 1991, *The Earth through Time*, fourth edition. Fortworth: Saunders Col. Publ., 651pp.

Li, Z.-X., Zhang, L., and Powell, C. McA., 1995, South China in Rodinia: Part of the missing link between Australia-East Antarctica and Laurentia? *Geology* **23**, 407–10.

Li, Z-X., Li, X.-H., Zhou, H., and Kinny, P. D., 2002, Grenvillian continental collision in south China: New SHRIMP U-Pb zircon results and implications for the configuration of Rodinia. *Geology* **30**, 163–6.

Link, P. K., Christie-Blick, N., Steward, J. H., Miller, J. M. G., Devlin, W. J., and Levy, M., 1993, Late Proterozoic strata of the United States cordillera. *In* Reed, J. C., Jr., Bickford, M. E., Houston, R. S., Link, P. K., Rankin, D. W., Sims, P. K., and Van Schmus, W. R., eds. *The Geology of North America*, vol. C-2, *Precambrian: Conterminous U. S.*, pp. 536–58. Boulder, CO: Geological Society of America.

Lipman, P. W., 1980, Cenozoic volcanism in the western United States: Implications for continental tectonics. *In* Burchfiel, B. C., Oliver, J. E., and Silver, L. T., eds., *Studies in Geophysics, Continental Tectonics*, pp. 161–74. Washington, D.C.: National Academy of Sciences.

1992, Magmatism in the Cordilleran United States; Progress and problems. *In* Burchfiel, B. C., Lipman, P. W., and Zoback, M. L., eds., *The Geology of North America*, vol. G-3, *The Cordilleran Orogen: Conterminous U. S.*, pp. 481–514. Boulder, CO: Geological Society of America.

Livaccari, R. F., 1991, Role of crustal thickening and extensional collapse in the tectonic evolution of the Sevier-Laramide orogeny, western United States. *Geology* **19**, 1104–07.

Livaccari, R. F., and Perry, F. V., 1993, Isotopic evidence for preservation of Cordilleran lithospheric mantle during the Sevier-Laramide orogeny, western United States. *Geology* **21**, 719–22.

Lockley, M., Matsukawa, M., and Hunt, A., no date, *Tracking Dinosaurs: New Interpretations of Dinosaurs based on Footprints*. Guidebook to Accompany An International Dinosaur Exhibit, University of Colorado at Denver, 68pp.

Lockley, M. G., 1998, The vertebrate track record. *Nature* **396**, 429–32.

Logan, B. W., 1987, The MacLeod evaporite basin, western Australia: Holocene environments, sediments, and geological evolution. Tulsa, OK: American Association of Petroleum Geologists, Memoir **44**, 140pp.

Lopez, R., Cameron, K. L., and Jones, N. W., 2001, Evidence for Paleoproterozoic, Grenvillian, and Pan-African age Gondwanan crust beneath northeastern Mexico. *Geol. Soc. Am. Bull.* **107**, 195–214.

Lucas, S. G., and Anderson, O. J., 1996, The Middle Jurassic Todilto salina basin, American Southwest. *In* Morales, M., ed., The Continental Jurassic. *Mus. North. Ariz., Bull.* **60**, 479–82.

1997, Lower Cretaceous stratigraphy on the Colorado Plateau. *N. Mex. Geol. Soc. Guidebook* **48**, 6–7.

Lucas, S. G., and Lawton, T. F., 2000, Stratigraphy of the Bisbee Group (Jurassic-Cretaceous), Little Hatchet Mountains, New Mexico. *N. Mex. Geol. Soc. Guidebook* **51**, 175–201.

Lucas, S. G., Williamson, T. E., Smith, L. N., Wright-Dunbar, R., Hallett, B., Kues, B. S., Hoffman, G., Hunt, A. P., Love, D. W., McLemore, V. T., and Hadley, R. F., 1992, First-day road log, from Cuba to La Ventana, San Luis, Cabezon, Mesa Portales, Mesa de Cuba and return to Cuba. *N. Mex. Geol. Soc. Guidebook* **43**, 1–32.

Lucas, S. G., Ziegler, K. E., Lawton, T. F., and Filkorn, H. F., 2001, Late Jurassic invertebrate fossils from the Little Hatchet Mountains, southwestern New Mexico. *N. Mex. Geol.* **23**, (1), 16–20.

Luedke, R. G., and Smith, R. L., 1978, Map showing distribution, composition, and age of late Cenozoic volcanic centers in Arizona and New Mexico. *U. S. Geol. Surv. Misc. Invest. Map* I-1091-A, 1:1,000,000.

Luhr, J. F., Henry, C. D., Housh, T. B., Aranda-Gómez, J. J., and McIntosh, W. C., 2001, Early extension and associated mafic alkalic volcanism from the southern Basin and Range province: Geology and petrology of the Rodeo and Nazas volcanic fields, Durango, Mexico. *Geol. Soc. Am. Bull.* **113**, 760–73.

Mack, G. H., Kottlowski, F. E., and Seager, W. R., 1998, The stratigraphy of south-central New Mexico. *N. Mex. Geol. Soc. Guidebook* **49**, 135–54.

MacLeod, K. G., Huber, B. T., and Fullagar, P. D., 2001, Evidence for a small ($\sim$0.000 030) but resolvable increase in seawater $^{87}Sr/^{86}Sr$ ratios across the Cretaceous-Tertiary boundary. *Geology* **29**, 303–6.

MacNiocaill, C., van der Pluijm, B. A., and Van der Voo, R., 1997, Ordovician paleogeography and the evolution of the Iapetus ocean. *Geology* **25**, 159–62.

Marzolf, J. E., 1988, Reconstruction of Late Triassic and Early and Middle Jurassic sedimentary basins: Southwestern Colorado Plateau to the eastern Mojave Desert. *In* Weide, D. L., and Faber, M. L., eds., *This Extended Land – Geological Journeys in the Southern Basin and Range*, Field Trip Guidebook, pp. 177–200. Las Vegas, NV: Dept of Geosciences, University of Nevada.

Mathis, R. S., 1982, Mid-Tertiary detachment faulting in the southeastern Newberry Mountains, Clark County, Nevada. *In* Frost, E. G., and Martin, D. L., eds., *Mesozoic-Cenozoic Tectonic Evolution of the Colorado River Region, California, Arizona, and Nevada*, pp. 327–40. San Diego, Cordilleran Publishers.

Mauger, R. L., McDowell, F. W., and Bount, J. G., 1983, Grenville-age Precambrian rocks of the Los Filtros area near Aldama, Chihuahua, Mexico. *In* Clark, K. F., and Goodell, P. C., eds., *Geology and Mineral Resources of North-Central Chihuahua*, pp. 165–8. El Paso. El Paso Geological Society.

McGookey, D. P., Haun, J. D., Hale, L. A., Goodell, H. G., McCubbin, D. G., Weimer, R. J., and Wulf, G. R., 1972, Cretaceous System. *In* Mallory, W. W., Mudge, M. R., Swanson, V. E., Stone, D. S., and Lumb, W. E., eds., *Geologic Atlas of the Rocky Mountain Region, United States of America*, pp. 190–228. Denver, CO: Rocky Mountain Association of Geologists.

McIntosh, W. C., and Bryan, C., 2000, Chronology and geochemistry of the Boot Heel volcanic field, New Mexico. *N. Mex. Geol. Soc. Guidebook* **51**, 157–74.

McLemore, V. T., McIntosh, W. C., and Hawley, J. W., 2000, First-day road log, from Lordsburg to Ruth Mine (Lordsburg District) to Twelve Mile Hill to Rock House Canyon (Pyramid Mountains) to Burgett's Greenhouses (Animas Valley) to Steins (Peloncillo Mountains), New Mexico. *N. Mex. Geol. Soc. Guidebook* **51**, 1–16.

McQuarrie, N., and Chase, C. G., 2000, Raising the Colorado Plateau. *Geology* **28**, 91–4.

Middleton, L. T., and Elliott, D. K., 1990, Tonto Group. *In* Beus, S. S., and Morales, M., eds., *Grand Canyon Geology*, pp. 83–106. Oxford University Press/Museum of Northern Arizona.

Middleton, L. T., Elliott, D. K., and Morales, M., 1990, Coconino Sandstone. *In* Beus, S. S., and Morales, M., eds., *Grand Canyon Geology*, pp. 183–202. Oxford University Press/Museum of Northern Arizona.

Miller, J. S., Glazner, A. F., Walker, J. D., and Martin, M. W., 1995, Geochronologic and isotopic evidence for Triassic-Jurassic emplacement of the eugeoclinal allochthon in the Mojave Desert region, California. *Geol. Soc. Am. Bull.* **107**, 1441–57.

Molenaar, C. M., 1983. Major depositional cycles and regional correlations of Upper Cretaceous rocks, southern Colorado Plateau and adjacent areas. *In* Reynolds,

M. W., and Dolly, E. D., eds., *Mesozoic Paleogeography of the West-central United States: Rocky Mountain Section*, Society of Economic Paleontologists and Mineralogists, Rocky Mountain Paleogeography, Symposium 2, pp. 201–24.

Morgan, J., Warner, M., and the Chicxulub Working Group, Brittan, J., Buffler, R., Camargo, A., Christeson, G., Denton, P., Hildebrand, A., Hobbs, R., Macintyre, H., Mackenzie, G., Maguire, P., Marin, L., Nakamura, Y., Pilkington, M., Sharpton, V., Snyder, D., Suarez, G., and Trejo, A., 1997, Size and morphology of the Chicxulub impact crater. *Nature* **390**, 472–6.

Mosher, S., 1993, Western extension of Grenville age rock; Texas. *In* Reed, J. C., Jr., Bickford, M. E., Houston, R. S., Link, P. K., Rankin, D. W., Sims, P. K., and Van Schmus, W. R., eds., *The Geology of North America*, vol. C-2, *Precambrian: Conterminous U. S.*, pp. 365–78. Boulder, CO: Geological Society of America.

1998, Tectonic evolution of the southern Laurentian Grenville orogenic belt. *Geol. Soc. Am. Bull.* **110**, 1357–75.

Mueller, P. A., Wooden, J. L., and Nutman, A. P., 1992, 3.96 Ga zircons from an Archean quartzite, Beartooth Mountains, Montana. *Geology* **20**, 327–30.

Musashi, M., Isozaki, Y., Koike, T., and Kreulen, R., 2001, Stable carbon isotope signature in mid-Panthalassa shallow-water carbonates across the Permo-Triassic boundary: evidence for ^{13}C-depleted superocean. *Earth Planet. Sci. Lett.* **191**, 9–20.

Mukhopadhyay, S., Farley, K. A., and Montanari, A., 2001, A short duration of the Cretaceous-Tertiary boundary event: evidence from extraterrestrial Helium-3. *Science* **291**, 1952–5.

Mulholland, J. W., 1998a, Sequence stratigraphy: Basic elements, concepts, and terminology. *The Leading Edge of Geophysics Exploration*, Society of Exploration Geophysicists, Jan., pp. 37–40.

1998b, Sequence architecture. *The Leading Edge of Geophysics Exploration*, Society of Exploration Geophysicists, June, pp. 767–71.

1998c, The Parasequence. *The Leading Edge of Geophysics Exploration*, Society of Exploration Geophysicists, October, pp. 1374–6.

Myers, D. A., 1982, Stratigraphic summary of Pennsylvanian and Lower Permian rocks, Manzano Mountains, New Mexico. *N. Mex. Geol. Soc. Guidebook* **33**, 233–7.

Nier, A. O., 1989, Some reminiscences of mass spectrometry and the Manhattan Project. *J. Chem. Educ.* **66**, 385–8.

Nieto-Samaniego, A. F., Ferrari, L., Alaniz-Alvarez, S. A., and Labarthe-Hernández, G., 1999, Variation of Cenozoic extension and volcanism across the southern Sierra Madre Occidental volcanic province, Mexico. *Geol. Soc. Am. Bull.* **111**, 347–63.

Nyman, M. W., Karlstrom, K. E., Kirby, E., and Graubard, C. M., 1994, Mesoprotero-zoic contractional orogeny in western North America: Evidence from ca. 1.4 Ga plutons. *Geology* **22**, 901–4.

Oldow, J. S., Bally, A. W., Avé Lallemant, H. G., and Leeman, W. P., 1989, Phanerozoic evolution of the North American Cordillera; United States and Canada. *In* Bally, A. W., and Palmer, A. R., eds., *The Geology of North America*, v. A, *The Geology of North America; An Overview*, pp. 139–232. Boulder, CO: Geological, Society of America.

Oldow, J. S., Aiken, C. L. V., Hare, J. L., Ferguson, J. F., and Hardyman, R. F., 2001, Active displacement transfer and differential block motion within the central Walker Lane, western Great Basin. *Geology* **29**, 19–22.

Pang, M., and Nummedal, D., 1995, Flexural subsidence and basement tectonics of the Cretaceous western interior basin, United States. *Geology* **23**, 173–6.

Park, J. K., 1994, Palaeomagnetic constraints on the position of Laurentia from middle Neoproterozoic to Early Cambrian times. *Precambrian Res.* **69**, 95–112.

Park, J. K., Buchan, K. L., and Harlan, S. S., 1995, A proposed giant radiating dyke swarm fragmented by the separation of Laurentia and Australia based on pa-leomagnetism of ca. 780 Ma mafic intrusions in western North America. *Earth Planet. Sci. Lett.* **132**, 129–39.

Parsons, B., and Sclater, J. G., 1977, An analysis of the variation of ocean floor bathymetry and heat flow with age. *J. Geophys. Res.* **82**, 803–27.

Perry, F. V., Baldridge, W. S., and DePaolo, D. J., 1987, Role of asthenosphere and lithosphere in the genesis of late Cenozoic basaltic rocks from the Rio Grande rift and adjacent regions of the southwestern United States. *J. Geophys. Res.* **92**, 9193–213.

Peterson, F., 1988, A synthesis of the Jurassic system in the southern Rocky Mountain region. *In* Sloss, L. L., ed., *The Geology of North America*, vol. D-2, *Sedimentary Cover – North American Craton*, pp. 65–76. Boulder. CO: Geological Society of America.

Philpotts, A. R., 1990, *Principles of Igneous and Metamorphic Petrology*. Englewood Cliffs, NJ: Prentice Hall, 498pp.

Phinney, E. J., Mann, P., Coffin, M. F., and Shipley, T. H., 1999, Sequence stratigra-phy, structure, and tectonic history of the southwestern Ontong Java Plateau adjacent to the North Solomon Trench and Solomon Islands arc. *J. Geophys. Res.* **104**, 20, 449–66.

Pillmore, C. L., and Flores, R. M., 1987, Stratigraphy and depositional environ-ments of the Cretaceous-Tertiary boundary clay and associated rock, Raton basin, New Mexico and Colorado. *In* Fassett, J. E., and Rigby, J. K. Jr, eds., *The*

Cretaceous-Tertiary Boundary in the San Juan and Raton Basins, pp. 111–30. Boulder, CO: Geological Society of America.

Pillmore, C. L., and Flores, R. M., 1990, Cretaceous and Paleocene rocks of the Raton basin, New Mexico and Colorado – Stratigraphic-environmental framework. *N. Mex. Geol. Soc. Guidebook* **41**, 333–6.

Piper, J. D. A., 2000, The Neoproterozoic supercontinent: Rodinia or Palaeopangaea? *Earth Planet. Sci. Lett.* **176**, 131–46.

Pittenger, M. A., Marsaglia, K. M., and Bickford, M. E., 1994, Depositional history of the Middle Proterozoic Castner Marble and basal Mundy Breccia, Franklin Mountains, west Texas. *J. Sediment. Res.* **B64**, 282–97.

Powell, C.McA., McElhinny, M. W., Meert, J. G., and Park, J. K., 1993, Paleomagnetic constraints on timing of the Neoproterozoic breakup of Rodinia and the Cambrian formation of Gondwana. *Geology* **21**, 889–92.

Prave, A. R., 1999, Two diamictites, two cap carbonates, two δ^{13}C excursions, two rifts: The Neoproterozoic Kingston Peak Formation, Death Valley, California. *Geology* **27**, 339–42.

Price, J. G., Henry, C. D., Parker, D. F., and Barker, D. S. (eds.), 1986, Igneous geology of Trans-Pecos Texas, Field Trip Guide and Research Articles. *Texas Bur. Econ. Geol. Guidebook* **23**, 360pp.

Prothero, D. R., and Schwab, F., 1996, *Sedimentary Geology. An Introduction to Sedimentary Rocks and Stratigraphy*. New York: W. H. Freeman and Company, 575pp.

Pysklywec, R. N., and Mitrovica, J. X., 1998, Mantle flow mechanisms for the large-scale subsidence of continental interiors. *Geology* **26**, 687–90.

Rämö, O. T., and Calzia, J. P., 1998, Nd isotopic composition of cratonic rocks in the southern Death Valley region: Evidence for a substantial Archean source component in Mojavia. *Geology* **26**, 891–4.

Rees, P. McA., 2002, Land-plant diversity and the end-Permian mass extinction. *Geology* **30**, 827–30.

Reneau, S. L., Gardner, J. N., and Forman, S. L., 1996, New evidence for the age of the youngest eruptions in the Valles caldera, New Mexico. *Geology* **24**, 7–10.

Retallack, G. J., 1996, Acid trauma at the Cretaceous-Tertiary boundary in eastern Montana. *GSA Today* **6**(5), 1–7.

Retallack, G. J., Veevers, J. J., and Morante, R., 1996, Global coal gap between Permian-Triassic extinction and Middle Triassic recovery of peat-forming plants. *Geol. Soc. Am. Bull.* **108**, 195–207.

Riggs, N. R., Lehman, T. M., Gehrels, G. E., and Dickinson, W. R., 1996, Detrital zircon link between headwaters and terminus of the Upper Triassic Chinle-Dockum paleoriver system. *Science* **273**, 97–100.

Rinehart, E. J., Sanford, A. R., and Ward, R. M., 1979, Geographic extent and shape of an extensive magma body at mid-crustal depths in the Rio Grande rift near Socorro, New Mexico. *In* Riecker, R. E., ed., *Rio Grande Rift: Tectonics and Magmatism, Am. Geophys. U., Spec. Publ.* pp. 237–51. Washington, D.C.: American Geophysical Union.

Rivers, T., and Corrigan, D., 2000, Convergent margin on southeastern Laurentia during the Mesoproterozoic: tectonic implications. *Cana. J. Earth Sci.* **37**, 359–83.

Rudnick, R. L., and Fountain, D. M., 1995, Nature and composition of the continental crust: A lower crustal perspective. *Rev. Geophys.* **33**, 267–309.

Ruiz, J., Patchett, P. J., and Ortega-Gutierrez, F., 1988, Proterozoic and Phanerozoic basement terranes of Mexico from Nd isotopic studies. *Geol. Soc. Am. Bull.* **100**, 274–81.

Sayre, W. O., Ort, M. H., and Graham, D., 1995, Capulin Volcano is approximately 59,100 years old: cosmogenic helium aging technique key to clearing up age old question. *Park Science*, **15**, 10–11.

Schlager, W., 1999, Sequence stratigraphy of carbonate rocks. *The Leading Edge of Geophysics Exploration*, August, 901–4, 906–7.

Schlue, J. W., Aster, R. C., and Meyer, R. P., 1996, A lower crustal extension to a midcrustal magma body in the Rio Grande Rift, New Mexico. *J. Geophys. Res.* **101**, 25, 283–91.

Schnabel, C., Pierazzo, E., Xue, S., Herzog, G. F., Masarik, J., Cresswell, R. G., di Tada, M. L., Liu, K., and Fifield, L. K., 1999, Shock melting of the Canyon Diablo impactor: Constraints from Nickel-59 contents and numerical modeling. *Science* **285**, 85–8.

Schultz, P. H., and D'Hondt, S., 1996, Cretaceous-Tertiary (Chicxulub) impact angle and its consequences. *Geology* **24**, 963–7.

Schwartz, H. L., and Gillette, D. D., 1994, Geology and taphonomy of the *Coelophysis* quarry, Upper Triassic Chinle Formation, Ghost Ranch, New Mexico. *J. Paleont.* **68**, 1118–30.

Schweickert, R. A., 1976, Early Mesozoic rifting and fragmentation of the Cordilleran orogen in the western USA. *Nature* **260**, 586–91.

Schweickert, R. A., and Cowan, D. S., 1975, Early Mesozoic tectonic evolution of the western Sierra Nevada, California. *Geol. Soc. Am. Bull.* **86**, 1329–36.

Sears, J. W., and Price, R. A., 2000, New look at the Siberian connection: No SWEAT. *Geology* **28**, 423–6.

Sheehan, P. M., and Fastovsky, D. E., 1992, Major extinctions of land-dwelling vertebrates at the Cretaceous-Tertiary boundary, eastern Montana. *Geology* **20**, 556–60.

Sigmarsson, O., Martin, H., and Knowles, J., 1998, Melting of a subducting oceanic crust from U-Th disequilibria in austral lavas. *Nature* **394**, 566–9.

Simkin, T., and Fiske, R. S., 1983, *Krakatau 1993: The Volcanic Eruption and its Effects*. Washington, D. C.: Smithsonian Institution Press, 464pp.

Sloss, L. L., 1965, Sequences in the cratonic interior of North America. *Geol. Soc. Am. Bull.* **74**, 93–114.

Smith, A. G., Smith, D. G., and Funnell, B. M., 1994, *Atlas of Mesozoic and Cenozoic Coastlines*. Cambridge: Cambridge University Press, 99pp.

Smith, R. M. H., and Ward, P. D., 2001, Pattern of vertebrate extinctions across an event bed at the Permian-Triassic boundary in the Karoo basin of South Africa. *Geology* **29**, 1147–50.

Snyder, D. B., and Hobbs, R. W., 1999, Ringed structural zones with deep roots formed by the Chicxulub impact. *J. Geophys. Res.* **104**, 10,743–55.

Soegaard, K., 1990, Fan-delta and braid-delta systems in Pennsylvanian Sandia Formation, Taos trough, northern New Mexico: Depositional and tectonic implications. *Geol. Soc. Am. Bull.* **102**, 1325–43.

Soegaard, K., and Caldwell, K. R., 1990, Depositional history and tectonic significance of alluvial sedimentation in the Permo-Pennsylvanian Sangre de Cristo Formation, Taos trough, New Mexico. *N. Mex. Geol. Soc. Guidebook* **41**, 277–89.

Soegaard, K., and Callahan, D. M., 1994, Late Middle Proterozoic Hazel Formation near Van Horn, Trans-Pecos Texas: Evidence for transpressive deformation in Grenvillian basement. *Geol. Soc. Am. Bull.* **106**, 413–23.

Soegaard, K., and Eriksson, K. A., 1985, Evidence of tide, storm, and wave interaction on a Precambrian siliciclastic shelf: the 1,700 m.y. Ortega Group, New Mexico. *J. Sed. Petrol.* **55**, 672–84.

Sonder, L. J., and Jones, C. H., 1999, Western United States extension: How the West was widened. *A. Rev. Earth Planet. Sci.* **27**, 417–62.

Stanesco, J. D., and Campbell, J. A., 1989, Eolian and noneolian facies of the Lower Permian Cedar Mesa Sandstone Member of the Cutler Formation, southeastern Utah. *U. S. Geol. Surv. Bull.* **1808**-F, F1–13.

Steck, L. K., Thurber, C. H., Fehler, M. C., Lutter, W. J., Roberts, P. M., Baldridge, W. S., Stafford, D. G., and Sessions, R., 1998, Crust and upper mantle *P* wave velocity structure beneath Valles caldera, New Mexico: Results from the Jemez teleseismic tomography experiment. *J. Geophys. Res.* **103**, 24,301–20.

Stevens, C. H., Stone, P., and Kistler, R. W., 1992, A speculative reconstruction of the Middle Paleozoic continental margin of southwestern North America. *Tectonics* **11**, 405–19.

Stevenson, G. M., and Baars, D. L., 1987, The Paradox: A pull-apart basin of Pennsylvanian age. *Four-Corners Geol. Soc. Guidebook* **10**, 31–50.

Stewart, J. H., Anderson, T. H., Haxel, G. B., Silver, L. T., and Wright, J. E., 1986, Late Triassic paleogeography of the southern Cordillera – the problem of a source for voluminous volcanic detritus in the Chinle Formation of the Colorado Plateau region. *Geology* **14**, 567–70.

Sutherland, P. K., and Harlow, F. H., 1973, Pennsylvanian brachiopods and biostratigraphy in southern Sangre de Cristo Mountains, New Mexico. *N. Mex. Bur. Mines Min. Res. Mem.* **27**, 173pp.

Swanson, E. R., and McDowell, F. W., 1984, Calderas of the Sierra Madre Occidental volcanic field, western Mexico. *J. Geophys. Res.* **89**, 8787–99.

Swisher, C. C., Grajalesnishimura, J. M., Montanari, A., Margolis, S. V., Claeys, P., Alvarez, W., Renne, P., Cedillopardo, E., and Maurrasse, F. J. M. R., 1992, Coeval Ar-40/Ar-39 ages of 65.0 million years ago from Chicxulub crater melt rock and Cretaceous-Tertiary boundary tektites. *Science* **257**, 954–8.

Tatsumi, Y., Shinjoe, H., Ishizuka, H., Sager, W. W., and Klaus, A., 1998, Geochemical evidence for a mid-Cretaceous superplume. *Geology* **26**, 151–4.

Thatcher, W., Foulger, G. R., Julian, B. R., Svarc, J., Quilty, E., and Bawden, G. W., 1999, Present-day deformation across the Basin and Range province, western United States. *Science* **283**, 1714–18.

Timmons, J. M., Karlstrom, K. E., Dehler, C. M., Geissman, J. W., and Heizler, M. T., 2001, Proterozoic multistage (ca. 1.1 and 0.8 Ga) extension recorded in the Grand Canyon Supergroup and establishment of northwest- and north-trending tectonic grains in the southwestern United States. *Geol. Soc. Am. Bull.* **113**, 163–80.

Tohver, E., van der Pluijm, B. A., Van der Voo, R., Rizzotto, G., and Scandolara, J. E., 2002, Paleogeography of the Amazon craton at 1.2 Ga: early Grenvillian collision with the Llano segment of Laurentia. *Earth Planet. Sci. Lett.* **199**, 185–200.

Troxel, B. W., 1974, Geologic guide to the Death Valley region, California and Nevada. *Geol. Soc. Am.*, Field Trip No. 1, 70th An. Mtg., Cordilleran Section, pp. 3–16. Shoshone, CA: Death Valley Publ. Co.

Trudgill, B., and Cartwright, J., 1994, Relay-ramp forms and normal fault linkages, Canyonlands National Park, Utah. *Geol. Soc. Am. Bull.* **106**, 1143–57.

Turner, C. E., 1990, Toroweap Formation. *In* Beus, S. S., and Morales, M., eds., *Grand Canyon Geology*, pp. 203–23. Oxford University Press.

Turner, C. E., and Fishman, N. S., 1991, Jurassic Lake T'oo'dichi': A large alkaline, saline lake, Morrison Formation, eastern Colorado Plateau. *Geol. Soc. Am. Bull.* **103**, 538–58.

Vajda, V., Raine, J. I., and Hollis, C. J., 2001, Indication of global deforestation at the Cretaceous-Tertiary boundary by New Zealand fern spike. *Science* **294**, 1700–2.

van der Lee, S., and Nolet, G., 1997, Seismic image of the subducted trailing fragments of the Farallon plate. *Nature* **386**, 266–9.

Van Schmus, W. R., and 22 others, 1993, Transcontinental Proterozoic provinces. *In* Reed, J. C., Jr., Bickford, M. E., Houston, R. S., Link, P. K., Rankin, D. W., Sims, P. K., and Van Schmus, W. R., eds., *The Geology of North America*, vol. C-2, *Precambrian: Conterminous U. S.*, pp. 171–334. Boulder, CO: Geological Society of America.

Walton, E. K., Randall, B. A. O., Battey, M. H., and Tomkeieff, O., eds., 1983, *Dictionary of Petrology*. New York: Wiley, 680pp.

Wang, K., Geldsetzer, H. H. J., and Krouse, H. R., 1994, Permian-Triassic extinction: Organic $\delta^{13}C$ evidence from British Columbia, Canada. *Geology* **22**, 580–4.

Ward, P. D., Montgomery, D. R., and Smith, R., 2000, Altered river morphology in South Africa related to the Permian-Triassic extinction. *Science* **289**, 1740–3.

Warme, J. E., and Kuehner, H.-C., 1998, Anatomy of an anomaly: The Devonian catastrophic Alamo impact breccia of southern Nevada. *In* Ernst, W. G., and Nelson, C. A., eds., *Integrated Earth and Environmental Evolution of the Southwestern United States*. The Clarence A. Hall, Jr., Volume, Geol. Soc. Am. International Book Series, pp. 80–107.

Warme, J. E., and Sandberg, C. A., 1996, Alamo megabreccia: Record of a Late Devonian impact in southern Nevada. *GSA Today* **6**(1), 1–7.

Watkinson, A. J., and Alexander, J. I. D., 1993, Castile microfolding. *N. Mex. Geol. Soc. Guidebook* **44**, 14–16.

Weil, A. B., Van der Voo, R., Mac Niocaill, C., and Meert, J. G., 1998, The Proterozoic supercontinent Rodinia: paleomagnetically derived reconstructions for 1100 to 800 Ma. *Earth Planet. Sci. Lett.* **154**, 13–24.

Weimer, P., and Slatt, R. M., 1999, Turbidite systems, Part 1: Sequence and seismic stratigraphy. *The Leading Edge of Geophysics Exploration*, pp. 454–63.

Wernicke, B., 1992, Cenozoic extensional tectonics of the U. S. Cordillera. *In* Burchfiel, B. C., Lipman, P. W., and Zoback, M. L., eds, *The Geology of North America*, vol. G-3, *The Cordilleran Orogen: Conterminous U. S.*, pp. 553–81. Boulder, CO: Geological Society of America.

Wernicke, B., and Snow, J. K., 1998, Cenozoic tectonism in the central Basin and Range: Motion of the Sierran – Great Valley Block. *Int. Geol. Rev.* **40**, 403–10.

Wernicke, B. P., Christiansen, R. L., England, P. C., and Sonder, L. J., 1987, Tectonomagmatic evolution of Cenozoic extension in the North American Cordillera. *In* Coward, M. P., Dewey, J. F., and Hancock, P. L., eds., *Continental Extensional Tectonics*, Geol. Soc. London, Spec. Publ. **28**, 203–21.

White, F. A., and Wood, G. M., 1968, *Mass Spectrometry: Applications in Science and Engineering*. New York: Wiley, 352pp.

Wilde, S. A., Valley, J. W., Peck, W. H., and Graham, C. M., 2001, Evidence from detrital zircons for the existence of continental crust and oceans on the Earth 4.4 Gyr ago. *Nature* **409**, 175–8.

Williams, E. G., Wright, L. A., and Troxel, B. W., 1974, The Noonday Dolomite and Equivalent stratigraphic units, southern Death Valley region, California. *Geol. Soc. Am., Field Trip No. 1*, 70th An. Mtg., Cordilleran Section, pp. 73–7. Shoshone, CA: Death Valley Publ. Co.

Williams-Stroud, S. C., 1994, Solution to the paradox? Results of some chemical equilibrium and mass balance calculations applied to the Paradox basin evaporite deposit. *Am. J. Sci.* **294**, 1189–28.

Wilson, J. L., 1989, Lower and Middle Pennsylvanian strata in the Orogrande and Pedregosa basins, New Mexico. *N. Mex. Bur. Mines Min. Res. Bull.* **124**, 16pp.

Wilson, P., Rais, J., Reigber, Ch., Reinhart, E., Ambrosius, B. A. C., Le Pichon, X., Kasser, M., Suharto, P., Majid, D. A., Yaakub, P., Almeda, R., and Boonphakdee, C., 1998, Study provides data on active plate tectonics in southeast Asia region. *Eos, Trans. Am. Geophys. Union* **79**, 545, 548–9.

Wingate, M. T. D., and Giddings, J. W., 2000, Age and palaeomagnetism of the Mundine Well dyke swarm, Western Australia: implications for an Australia-Laurentia connection at 755 Ma. *Precambrian Res.* **100**, 335–57.

Winkler, D. A., Jacobs, L. L., Congleton, J. D., and Downs, W. R., 1991, Life in a sand sea: Biota from Jurassic interdunes. *Geology* **19**, 889–92.

Wolbach, W. S., Anders, E., and Nazarov, M. A., 1990, Fires at the K/T boundary: Carbon at the Sumbar, Turkemia, site. *Geochim. Cosmochim. Acta* **54**, 1133–46.

Wolfe, J. A., Schorn, H. E., Forest, C. E., and Molnar, P., 1997, Paleobotanical evidence for high altitudes in Nevada during the Miocene. *Science* **276**, 1672–5.

Wolfe, J. A., Forest, C. E., and Molnar, P., 1998, Paleobotanical evidence of Eocene and Oligocene paleoaltitudes in midlatitude western North America. *Geol. Soc. Am. Bull.* **110**, 664–78.

Wooden, J. L., and Miller, D. M., 1990, Chronologic and isotopic framework for Early Proterozoic crustal evolution in the eastern Mojave Desert region, SE California. *J. Geophys. Res.* **95**, 20,133–46.

Wright, L. A., and Prave, A. R., 1993, Proterozoic-early Cambrian tectonostratigraphic record in the Death Valley region, California-Nevada. *In* Reed, J. C., Jr., Bickford, M. E., Houston, R. S., Link, P. K., Rankin, D. W., Sims, P. K., and Van Schmus, W. R., eds., *The Geology of North America*, vol. C-2, *Precambrian: Conterminous U. S.*, pp. 529–533. Boulder, CO: Geological Society of America.

Wright-Dunbar, R., 1992, Shoreline cyclicity and the transgressive record: A model based on Point Lookout Sandstone exposures at San Luis, New Mexico. *N. Mex. Geol. Soc. Guidebook* **43**, 12–14.

Wright-Dunbar, R., Zech, R. S., Crandall, G. A., and Katzman, D., 1992, Strandplain and deltaic depositional models for the Point Lookout Sandstone, San Juan basin and Four Corners platform, New Mexico and Colorado. *N. Mex. Geol. Soc. Guidebook* **43**, 199–206.

Wrucke, C. T., 1989, The Middle Proterozoic Apache Group, Troy Quartzite, and associated diabase of Arizona. *In* Jenney, J. P., and Reynolds, S. J., eds., *Geologic Evolution of Arizona, Ariz. Geol. Soc. Digest* 17, pp. 239–58.

1993, The Apache Group, Troy Quartzite, and Diabase: Middle Proterozoic rocks of central and southern Arizona. *In* Reed, J. C., Jr., Bickford, M. E., Houston, R. S., Link, P. K., Rankin, D. W., Sims, P. K., and Van Schmus, W. R., eds., *The Geology of North America*, vol. C-2, *Precambrian: Conterminous U. S.*, pp. 517–21. Boulder, CO: Geological Society of America.

Ye, H., Royden, L., Burchfiel, C., and Schuepbach, M., 1996, Late Paleozoic deformation of interior North America: The greater Ancestral Rocky Mountains. *Am. Assoc. Petrol. Geologists Bull.* **80**, 1397–432.

Yin, A., and Ingersoll, R. V., 1997, A model for evolution of Laramide axial basin in the southern Rocky Mountains, U. S. A. *Int. Geol. Rev.* **39**, 1113–23.

Young, G. M., 1995, Are Neoproterozoic glacial deposits preserved on the margins of Laurentia related to the fragmentation of two supercontinents? *Geology* **23**, 153–6.

Index